Cultivating Community

How discourse shapes the philosophy,
practice and policy of water management
in the Murray–Darling Basin

Amanda Shankland

PUBLIC AND SOCIAL POLICY SERIES

Gaby Ramia, Series Editor

The Public and Social Policy series publishes books that pose challenging questions about policy from national, comparative and international perspectives. The series explores policy design, implementation and evaluation; the politics of policy making; and analyses of particular areas of public and social policy.

Cultivating Community

How discourse shapes the philosophy,
practice and policy of water management
in the Murray–Darling Basin

Amanda Shankland

SYDNEY UNIVERSITY PRESS

First published by Sydney University Press
© Amanda Shankland 2024
© Sydney University Press 2024

Sydney University Press
Gadigal Country
Fisher Library F03
University of Sydney NSW 2006
Australia
sup.info@sydney.edu.au
sydneyuniversitypress.com

A catalogue record for this book is available from the National Library of Australia.

ISBN 9781743329771 paperback
ISBN 9781743329788 epub
ISBN 9781743329863 pdf

Cover design: Naomi van Groll
Cover image: Mouth of the Murray River. Photo: Amanda Shankland

We acknowledge the traditional owners of the lands on which Sydney University Press is located, the Gadigal people of the Eora Nation, and we pay our respects to the knowledge embedded forever within the Aboriginal Custodianship of Country.

This work is dedicated to my ancestors and descendants; may we seek to preserve the past and honour the future in all that we do.

Contents

Acknowledgements

This book was made possible with the support of many people. First, I want to thank my dissertation supervisor, Dr Peter Andrée, for the countless hours spent reviewing my work and the care, passion and attention to detail that helped this project come to life. I also owe thanks to my committee members, Marie-Josée Massicotte and Vandna Bhatia, for their extensive comments and suggestions that helped shape this book. I also want to thank Dr Michael Classens, Dr Randall Germain and Dr Arne Kislenko for offering me much needed support, friendship and encouragement along the way.

This work was made possible with the generous support of my Australian hosts, in particular Dr Vaughan Higgins and the Charles Sturt University community. Thank you also to Kathleen Bowmer, Ken Jury, and Tony and Elizabeth Wennerbom for your gracious hospitality in Australia. Additional thanks to Kathleen Bowmer, who has a long history of working on water issues in the Murray–Darling Basin, for providing access to all the documentation she had saved from meetings with government representatives, farmers and other actors.

This research was funded in part through the Australian Endeavour Scholarship, which allows international researchers to participate in research in Australia. Generous funding was also received from the Ontario Graduate Scholarship, the Isabel Bader Travel Award, and the Carleton Department of Political Science PhD Scholarship.

Thank you to the government officials in Canberra for their willingness to share their knowledge and experience with me. Special thanks to all the farmers who took time out of their busy schedules to meet with me – and who labour every day, rain or shine, to feed the world.

Thank you to all my family and my community who have stood by me and encouraged me through this long and tenuous process. Special thanks to my children, Idris, Ayman and Saidi, for sacrificing many a weekend with mom so that this project could come to fruition.

Significant events and developments

1979   Dartmouth Dam in Victoria is completed, becoming the basin's largest water storage

1985   First meeting of the Murray–Darling Basin Ministerial Council

1987   Murray–Darling Basin Agreement first signed, initially as an amendment to the River Murray Waters Agreement of 1914

1988   Establishment of the Murray–Darling Basin Commission

1992   New Murray–Darling Basin Agreement replaces River Murray Water Agreement

1996   Cap on diversions introduced for New South Wales, Victoria and South Australia

1997   Beginning of longest drought in Australia's recorded history (the Millennium Drought)

2000   Murray mouth closes and requires dredging

2000   National Action Plan for Salinity and Water Quality agreed

2002   *Living Murray* discussion paper released

2004   Intergovernmental Agreement on a National Water Initiative reached

2007   *Water Act 2007* introduced

2012   Murray–Darling Basin Plan becomes law

Map of the Murray–Darling Basin. Courtesy of Murray–Darling Basin Authority.

Measurements

gigalitre (GL)	1,000 megalitres or 1 billion litres
megalitre (ML)	1 million litres or the equivalent of a swimming pool that is 50 metres long by 20 metres wide and 1 metre deep

Abbreviations

ABC	Australian Broadcasting Corporation
ACF	Australian Conservation Foundation
ACT	Australian Capital Territory
CMA	Catchment Management Authority
CMB	Catchment Management Board
COAG	Council of Australian Governments
CSIRO	Commonwealth Scientific and Industrial Research Organisation
DLWC	NSW Department of Land and Water Conservation
DSNR	NSW Department of Sustainability and Natural Resources (formerly DLWC)
EPA	Environment Protection Authority
NAP	National Action Plan for Salinity and Water Quality
Ramsar Convention	Ramsar Convention on Wetlands of International Importance Especially as Waterfowl Habitat

List of Tables

Prologue

Communities in crisis

In July 2016, I arrived in Sydney. Two days later, I rented a car and was driving to Wagga Wagga, a small city located in the heart of New South Wales. I was alone in a land wholly unfamiliar to me. My satellite navigation system indicated I was an hour away from Wagga, but I had not seen another car on the road for more than an hour. I worried I had taken the wrong road. It seemed impossible that a populated city was close by. The empty landscape was dotted with sheep happily grazing the pastures. Aside from the vast distances and the absence of people, this was a familiar scene. The eucalyptus trees that sporadically appeared were the only reminders of a forgotten era before the arrival of the settlers who radically transformed the landscape, clearing the land for pasture so that it might resemble the English countryside.

A drive towards progress, individual freedom and dominion over nature is illustrated by the modernisation projects of the 20th century that have changed the face of the Earth. The Three Gorges Dam in China, for example, is one of the few projects so enormous that it can be seen by the naked eye from space. Massive hydrological projects that began in the 20th century helped to develop some of the most productive agricultural lands in the world. Such projects include the Hoover in the United States of America, the Kariba in Zambia, the Bhakra in India, the Aswan in Egypt, the W.A.C. Bennett in Canada and the Burrinjuck in Australia, to name a few.[1] These projects radically transformed ecological landscapes, damaging and destroying wetlands, and displacing people and communities to make farming possible. In time, productive and vibrant farming communities grew up around these dams and their contiguous irrigation and drainage systems.

Today, climate change, characterised by the increasing frequency of floods and droughts, threatens these communities. The effects of climate change also highlight the need to restore wetlands that act as vital carbon sinks. Retaining ecological sites like wetlands while ensuring the long-term sustainability of farm communities and their contributions to local and regional food security are significant challenges. These complex dynamics are exemplified within the Murray–Darling Basin, threatening its viability as an ecosystem and productive space. The Millennium Drought, which began in 1997 and lasted until 2009, decimated farming communities and highlighted the vulnerability of marshlands in the face of climate change.

The modernisation projects of the 20th century radically redefined natural landscapes. These projects were led by the state and relied on expert knowledge and bureaucratic planning. In Australia, with nearly all states having a history of starting as convict settlements run by colonial Britain, the state had a long history of leading development projects. These projects intensified dramatically in the 20th century as the state hoped to transform the landscape in ways that could make European-style agriculture possible.

Having never set foot in Australia, I developed an interest in bureaucratic planning and the role of the state in development during my time working at the Canadian Department of Agriculture. There, I witnessed how the knowledge of experts, scientists and bureaucrats was prioritised over farmers' knowledge. Further, some farmers' voices were given more consideration than others. For instance, some farmer organisations were removed from consultation lists because they were not considered "cooperative" enough. I was concerned that the apparent silencing of dissenting opinions would invariably negatively affect policy development. In this project, I hoped to conduct research that spoke to the needs of farming communities and develop a bottom-up engagement process. The Murray–Darling Basin offered an ideal

---

1   See Hiltzik 2010; Matanzima 2022; Prentice 1998; Rangachari 2006.

case study of a state-led development project that had received significant pushback from local farming communities.

Through my research, I came to recognise how discourses shape the ways we understand the natural world, how they empower and disempower different voices, and how they affect responses to environmental problems. Government officials, Indigenous communities, farmers and environmental advocates have distinct ways of understanding the world, as evidenced through discursive practices like the language and symbols each group uses. Discourses also define the parameters of what is considered acceptable and desirable. Uncovering the assumptions embedded in discourse is essential for understanding how interests are defined and how they can be redefined. For example, according to many Indigenous groups around the world, the Earth is conceived of as a mother who gives life and provides. On the other hand, government officials may generally understand the Earth as a resource that requires management. Such diverse conceptualisations frame definitions of problems and the policy choices that follow. Uncovering assumptions embedded in discourses can reveal alternative solutions to the problems associated with retaining vital ecological sites like wetlands while sustaining farm communities. A central precondition for reciprocal dialogue between communities and expert planners is the openness to alternative ideas, perspectives and knowledge.[2] Through open dialogue, discourses evolve, and new ideas can emerge. Overcoming hierarchal governance structures and top-down decision-making requires understanding how discourses shape assumptions about the world and the possibilities for change.

The dominant environmental discourses in Australia were moulded by the country's colonial past. Settlers had a sense of superiority and a drive to "improve" the land. This sense of superiority is rooted in monotheistic traditions, which teach believers that they are moulded in God's image, separate from, and holding dominion over, animals and nature. All civilisations are grounded in different ontological symbols that provide a collective identity or a "cosmion". Our symbols of the cosmion provide the internal structure – the symbol of Christ is an

---

2   Dryzek 2013; Hajer and Versteeg 2005; Litfin 1994.

example of this.[3] With Copernicus and the discovery that the Earth revolves around the Sun, Europeans recognised they were not at the centre of the universe. However, as technology and science challenged notions of God, the human-centred view of the world did not fade. Instead, people began to see science and technology as the primary means to control and direct the course of the natural world; this helped fuel the human desire for progress.

Rooted in liberal traditions, as exemplified in the writings of Enlightenment thinkers like Locke and Smith, is the idea that the value of nature is conferred only through the application of human labour, and unimproved nature is without value.[4] Emphasis on "progress" and individual land rights as central to a liberal economy has supported the commonly held view of land as a commodity that can be separated into parcels and partitioned from the surrounding environments. As cited in McCarty and Prudham, modern liberal thinking has dramatically restructured our relationship with nature through what Polanyi called "fictitious commodities".[5]

Enlightenment conceptions of individual freedom are also critical to our notions of "progress" and development. The Canadian philosopher George Grant argued that the most significant myth in our society is the myth of progress, because it depends on the assertion of absolute human sovereignty. He argued that people believe in the myth of progress to give meaning to their lives. Further, people believe in freedom as the absolute assertion of the self. However, when we experience personal freedom absolutely, we are no longer able to connect with the world around us. Grant asserted that the separation of myth (systems of meaning) from freedom (as realised only through the act of asserting oneself) can lead to what he describes as a type of schizophrenia. He calls this solipsism: the inability to conceive of others as truly human – they are instead objects. The language of the self makes us think we are absolute and responsible for our destinies. We come to believe that all our successes are determined by us alone, cutting the role of the community out of the picture.[6]

---

3   Voegelin 1952.
4   Voegelin 1952.
5   McCarty and Prudham 2004, 277, 281.
6   Grant 1998, 391–2.

It follows that in a world of automation, large cities and an unquestioned belief in progress and individual freedom, communities are disappearing. Like the city people, the farmers have become professionals, depending on technology, automation and growing their businesses.[7] Economic expansion through control of nature by science has become a common driving force of modern societies. The cost of such an approach has been, among other things, the disintegration of farm communities. The success of farmers is considered to rest squarely on the backs of individual farmers.

The environmental crisis is rooted in a crisis of human culture characterised by Western notions of progress and individual freedom. There is a common belief that progress is inevitable, that nature can be controlled and shaped to our individual desires and that there is little to learn from the past. These views have helped shape the nature of environmental discourse and governance. But some have challenged these views of the world and asked people to value nature as an end in itself and not just a means to satisfy human desires. As Hinchman and Hinchman explained, "what 'justifies' each individual being is not its potential serviceability for human schemes, but its irreplaceable contribution to the flourishing of the whole, a totality that *includes* human life and purposes but is not *defined* by them".[8] Culture itself should be treated as an expression of natural relationships. The ideas, metaphors and institutions we create are rooted in our relationship with the natural world. For example, the monoculture model of agriculture is symptomatic of a broader culture that has been shaped by the homogenising influence of technological modernity. There is a need to conceptualise society in ways that emphasise the interrelations between culture and nature, which are largely ignored in modern liberal regimes.[9]

Australian Indigenous peoples understand the relationship between culture and nature as inseparable. They understand that one element cannot be separated from the whole; the entire community of plants, animals and people must be accounted for. At the beginning

---

7   Grant 1998, 51.
8   Hinchman and Hinchman 1989, 210 (italics in original).
9   Hinchman and Hinchman 1989, 203, 214.

of colonialist conquests into Australia, early explorers learned about the Indigenous worldviews and their associated practices. But, over a short period, the pervading opinion was that little was to be learned from the Indigenous peoples and land should be fashioned by Western constructions of nature. The colonialist mindset was to work the land in ways that would achieve the desired results in the shortest time. This approach meant the settlers often destroyed the natural systems needed to sustain their production goals. By the mid-20th century, it became evident that there could be no environmental justice without an accounting of whole communities that are affected by changes to the landscape. This worldview is in stark contrast to that of the Western settlers who colonised the landscape and carried with them a vision of the world rooted in individual freedom and dominion over nature.

The 17th-century English philosopher Thomas Hobbes credited government institutions with delivering humanity from the state of nature.[10] This view has profoundly shaped interactions with the natural world as governments, corporations and institutions tend to see nature as separate from people. A discourse that prioritises people over nature has diminished our capacity to see the extent to which human activity has environmental consequences *and* to which environments exert influence on human affairs. The environmental movement is a healthy reaction to anthropocentrism, but it has often failed to recognise the well-entrenched and fragile ecological relationships between people and the environment resulting from our long historical legacy of interference in natural systems. In *The End of Nature*, Bill McKibbin explained how virtually no part of nature remains pristine. There is an increasing recognition that virtually no space remains untouched by human development. The environmental movement in Australia has historically been characterised by a protectionist ethic that seeks to maintain the "natural" condition of environmental spaces. Environmentalists worked to create spaces that were free from human development and interventions. But Indigenous communities tend to see people as part of natural spaces and focus on their role as caretakers. Today, the environmental movement is evolving in ways that consider

---

10   Hobbes 1651.

the critical role that people can play in caring for the environment in both productive and non-productive spaces.

If we hope to confront the enormous environmental challenges of our time, a dramatic shift in awareness, more closely aligned to the conceptions of nature of Australian Indigenous peoples, is needed: a view that recognises that people are a part of the natural world. Recognising the separation of people from their communities is crucial in understanding the disconnect people experience with the natural world.

Culture, the sets of practices and beliefs developed through human communities, determines how we interact with the land. Culture provides a set of instructions about how to live on the land. How we treat the land is representative of the health of human cultures. Human culture is a way to metabolise life. Through music, dance, conversation, telling stories and eating together, we create the communities that allow human culture to flourish. Human social relationships are thus a significant determinant of environmental outcomes. Murray Bookchin, a leading historical figure in the environmental movement, argued that the roots of ecological problems are closely tied to human social problems and can be solved by reorganising society along more ethical lines. Bookchin's approach acknowledged the co-dependent relationships between human communities and natural systems. Bookchin wrote: "ecological degradation, is in large part, a product of the degradation of human beings by hunger, material insecurity, class rule, hierarchal domination, patriarchy, ethnic discrimination, and competition".[10] Environmental health depends on the health of human cultures and communities. Low-income, minority and rural populations are disproportionately vulnerable to the effects of environmental destruction. Further, adaptation and mitigation in these communities are affected by financial, social and systemic constraints.

While seeking to gain a livelihood from the land, many farmers also see themselves as caretakers. They work to maintain natural spaces on their farms, reduce the use of fertilisers and pesticides, protect endangered species of insects, birds and amphibians, and farm in ways that account for the natural ecology of their land. This is the case

---

10   Hobbes 1651.

with many of the farmers living in the Murray–Darling Basin. Without a broader conception of ecological communities, which includes humans, plants and animals, we narrow the framing of environmental problems and limit potential solutions. This book draws attention to the vital role of community in achieving environmental outcomes. A shift in the discourse will require a recognition of the deep connections between nature and human communities.

From the outset, this research sought to explore what a bottom-up approach to addressing environmental problems might look like. My research focused on the role of farmers and the communities that support them, challenging a historically top-down approach by government. Discourse analysis was a fitting approach as it requires a high level of immersion in communities to uncover how respondents understand their circumstances and how they might effect change. During my five months of research in Australia, I drove more than 20,000 kilometres. The distance between farms made it nearly impossible to do more than one interview per day. My trips began at dawn, so I could arrive at my destination before nine in the morning and complete the interview before noon. Then I would have to drive an average of two to three hours to sleep at my house in Wagga Wagga or a hostel in one of the surrounding towns, often Griffith. Driving after dark was simply not possible as there were animals on the road, and a breakdown in the country could be dangerous. Even in a developed country like Australia, researchers face numerous hazards they cannot predict. Further, there can be a strong sense of isolation and even fear, fear I felt acutely when forced to sleep in my car one night. But visiting people in the spaces they are familiar with and meeting with them individually are essential for generating trust and creating an environment where people can speak freely about their experiences and opinions. Farmers appreciated that I made an effort to see them in such remote areas and that I had the proper footwear to get in the tractor or truck with them and continue the interview in the field.

For most farmers, this was the first time anyone had visited their farm to interview them, and they were happy to spend time with me and tell their stories. In many ways, there is an emotional aspect to this type of research because the researcher becomes invested in the outcome. One farmer, for example, lost her home and much of her land in flooding along the river shortly after my interview. As an outsider, it is difficult to comprehend how precarious the situation can be for farmers. When people talked about bankruptcy, divorce, drug addiction and even suicide in their communities in the wake of the drought, I could sense the despair. These experiences led me to want to contribute to a research project sensitive to the human costs of drought and government policy. For almost a year after the research, while I transcribed the interviews and worked in my teaching jobs, I felt a strong sense of despondency about my potential contribution. I experienced the fear that comes from understanding that a researcher's words and choices can genuinely affect the communities they undertake research in. The close relationships I developed with farmers meant that I heard strong and detailed arguments from them, which invariably affected my understanding of the situation. These relationships meant that I was committed to this group's wellbeing. This has resulted in a book that speaks more to the farmers' side of the story than other points of view. In hopes of presenting both sides of the issue, I have also seriously considered the perspectives of academic experts and policy analysts. In doing so, I hope that policymakers, particularly those at the Murray–Darling Basin Authority, will consider this account of events as a unique and critical perspective based on the observations of someone largely outside the problem. Even though researchers are expected to remain objective, I do not believe this is possible when you work closely with a population. Researchers are not unsympathetic observers; I had become emotionally invested in the outcome and hoped that my research would make a difference.

Introduction

In the language of the Wiradjuri people of central New South Wales, "Murrumbidgee" means "big boss". The Murrumbidgee River in the Murray–Darling Basin is so named because it dictates the way of life for those who live along it and depend on it. People here must learn to follow the uneasy ebb and flow of an ever-changing and largely unpredictable system. Living along the Murrumbidgee, Darling (Baaka) and Murray rivers of south-eastern Australia sustainably depends not only on how well one can predict the weather and ensuing conditions of the river *but* also on anticipating that some things are unpredictable. Life depends, to some degree, on the dictates of the river. This condition means that adaptation is an essential characteristic of survival. For the farmers, like for the Wiradjuri people, this is the reality that dictates the possibility of a future.

Currently, nearly 80 per cent of the world's population is threatened by an insecure water supply, and the vast majority of fresh water is dramatically affected by human activity. There is a political, social and environmental imperative to manage water sustainably. Australia, the driest inhabited continent, subject to extreme temporal and spatial variation in rainfall, faces significant challenges to its freshwater systems.[1] Maintaining the Murray–Darling Basin, Australia's most extensive freshwater system, is critical.

The Murray–Darling Basin is more than a million square kilometres in area and crosses four states and one territory. It includes some 77,000 kilometres of rivers and creeks and 30,000 wetlands. Despite its immense size, the river system has a modest average inflow.[2] The basin is essential to Australian agricultural interests and the rural communities supported by agriculture since colonisation. From 2011–12, the gross agricultural production in the Murray–Darling Basin was $19 billion, or around 40 per cent of the total Australian value of agricultural commodities. The catchment also provides water to some 2 million people.[3] Despite measures to conserve water, the Murray–Darling Basin is drying up, and so are the farm businesses that depend on it. Though there have been significant measures to conserve water, nearly half of the farmers in some parts of the Murray–Darling Basin have sold their water allocations back to the government, abandoning their cultivation of irrigation-dependent crops like table grapes and rice.

Farmers are integral to water conservation efforts in the basin, contributing a unique perspective rooted in their long history of adaptation efforts. But, a review of the history suggests that farmers have had difficulty influencing discussions around water management. By treating them primarily as a cause of water scarcity, governments often fail to recognise the potential contributions of farmers to addressing the crisis. Over their long history, negotiations concerning the management of the Murray–Darling Basin have frequently been top-down and have not produced the results desired by any of the actors involved. In some instances, initiatives also created distrust within communities and contributed to the crisis that actors were trying to mitigate.[4] For example, farmers were deeply concerned that the Commonwealth government would forcibly strip their water entitlements after they introduced plans in the *Water Act* of 2007 to

1   Swirepik, Burns et al. 2015.
2   Swirepik, Burns et al. 2015.
3   Chenoweth and Malano 2001.
4   Harley, Metcalf and Irwin 2014.

retrieve millions of litres of water from the system for environmental purposes. Some analysts argued that the causes of water mismanagement were found in federal interventions, while others said that mismanagement had happened primarily at the state level.[5] Either way, a hurried water reform process meant to divert water from production for environmental purposes occurred with limited input from farmers, and the ecological benefits remain questionable.[6] One clear result is that the process devastated farmers and communities implicated in these reforms.

Farmers have a wealth of local knowledge regarding water management on their farms and can make significant contributions in terms of solutions. But how is that knowledge considered by policy experts, and to what effect? The Murray–Darling Basin case reveals how farmers are included in policymaking and implementation processes meant to respond to challenging environmental circumstances. This book explores the impact of farmer knowledge and perspectives on water management discourse in the Murray–Darling Basin. I approach this inquiry by situating the knowledge and views of farmers within the broader policy discourse of water management in the basin.

This book addresses how environmental discourses shape the parameters of acceptable policy choices in the Murray–Darling Basin and subsequent outcomes. This is done by examining a series of questions: What are the defining discourses of water management in the Murray–Darling Basin, and how have some discourses gained authority over others? What forms of knowledge do these discourses legitimise? How have these discourses defined public policy historically and today? What difference do these discourses make to how land and water are managed? What alternative perspectives, knowledge and policy options are excluded, *and* what would be the policy implications of these alternative perspectives?

In response to these questions, this work identifies five environmental discourses in the farming and policy communities in the Murray–Darling Basin: administrative rationalism, economic rationalism, democratic pragmatism, green environmentalism and

---

5   Doyle and Kellow 1995.
6   Doyle and Kellow 1995; Lee and Ancev 2009.

community-centrism. It examines the origins of and assumptions embedded within these discourses. Further, it looks at how farmers influence these discourses and how the discourses affect farmers. Australian academic and environmental writer John Dryzek has previously discussed the first three discourses.[7] The fourth, green environmentalism, was a discourse I identified through my research. Green environmentalism is a dominant alternative discourse of environmental problem-solving in Australia. This often biocentric discourse is needed to understand environmental management decisions in the Murray–Darling Basin. I argue that the first four discourses played critical roles in shaping the parameters of acceptable policy choices in the Murray–Darling Basin from the 1950s to 2017, the period covered in this study. The final discourse, community-centrism, is one of resistance that has had a less direct effect on policy to date but has much to offer in terms of defining an alternative future for water management. This discourse was constructed through my observations of farmers' environmental management experiences in the Murray–Darling Basin, what they shared with me in interviews and a careful review of alternative environmental discourses.

John Dryzek identified three dominant discourses that Western societies have tended to work within when responding to environmental problems. He called these the "discourses of environmental problem solving": administrative rationalism, economic rationalism, and democratic pragmatism. Each of these discourses appears to be highly relevant in the case of the Murray–Darling Basin. Elements of each have played roles in defining policy choices in the basin in recent decades. The history of water management within the Murray–Darling Basin can be characterised mainly by administrative rationalism as policy design and implementation have largely been top-down, emphasising the expertise of scientists and bureaucratic control. Administrative rationalism is associated with professional resource management bureaucracies, central agencies, regulatory policy instruments, expert advisory commissions and rationalist policy-analysis techniques.[8] Administrative rationalism emphasises the

---

7   Dryzek 2013.
8   Dryzek 2013, 73, 75–98.

expert's role while downplaying the citizen participation role in building capacity for problem-solving. It has the goal of rapid modernisation under the guidance of those deemed expert authorities by the state, and it frequently assumes that nature is subordinate to human problem-solving. One fundamental problem with administrative rationalism is that it presents a false image of specific knowledge and benign power.[9] Further, powerful interests often interfere with decision-making. Decisions are guided by the interests or policy objectives of specific actors at the expense of satisfying the interests and goals of others.

Like administrative rationalism, economic rationalism also led to a myopic view of specific problems in the basin. In contrast to the centralising tendencies of administrative rationalism, economic rationalism is grounded in the notion that decision-making should happen at the individual level. Central to this discourse is the idea that individualism promotes competition, allocates resources more efficiently and thus contributes to positive economic growth.[10] Economic rationalism assumes that free markets are the best method of decentralising environmental planning and is often touted as the most reliable mechanism for dividing common resources. In the Murray–Darling Basin, adopting economic rationalism led to an emphasis on economic instruments to resolve water over-allocation problems, meaning that other solutions were often overlooked.

Economic rationalists argue that free markets and the protection of individual property rights are best for preventing what they view as the disasters associated with state-centric environmental planning. A consequence of this approach in the Murray–Darling Basin is that private rights regimes necessitated increased government intervention to address negative externalities. Proponents have argued that economic rationalism avoids the tragedy of the commons because property owners are more likely to care for private than public (or common) property.[11] But managing collective resources under private rights regimes is quite challenging, mainly because the level of

9 Torgerson and Paehlke 2005.
10 Dryzek 2013, 122–34.
11 Hardin 1968.

management intervention often elevates the need for government involvement.[12] As will be argued, this situation has often occurred in the Murray–Darling Basin. Economic rationalism also undervalues the multifunctionality of ecosystems by focusing only on economic outcomes in the short term.[13]

Dryzek's third primary environmental problem-solving discourse, democratic pragmatism, emphasises the practical application of ideas through democratic processes, such as environmental consultations and involving members of the broader public in consensus-building initiatives, rather than through the imposition of ideological force. Democratic pragmatism assumes that participants are informed and that special interests will not dominate. But, as we will see, in practice, the discourse of democratic pragmatism can reinforce the status quo and ignore the wider-scale social processes in which specific environmental issues are embedded.[14] Though the Commonwealth government was committed to consultation in the Murray–Darling Basin, farmers felt excluded for many reasons. For instance, they often thought that the government only initiated consultations after making decisions, that government representatives were unwilling to meet with farmers in the spaces they were accustomed to and that information was often inaccessible or incomprehensible. This research shows that attempts to democratise processes for managing the Murray–Darling Basin often failed and therefore had limited problem-solving potential. I argue that efforts towards democratic pragmatism were constrained by the overarching discourse of administrative rationalism and its impact on decision-making.

Each of the three discourses presents a different story of environmental water management. Together, they help unpack what happened in the Murray–Darling Basin. First, a legacy of administrative rationalism has shaped policy developments in the region and continues to do so. The increased influence of economic

---

12 Robertson 2007.
13 In agriculture, multifunctionality refers to the numerous benefits that agricultural may provide, generally to the non-trade benefits of agriculture (OECD 2001); Hollander 2007.
14 Dryzek 2013.

rationalism has complicated that story since the 1980s. Economic rationalism was manifest to the extent that it fits within the overall structure of administrative rationalism. Democratic pragmatism also informed problem-solving in the Murray–Darling Basin but primarily to support the status quo that was already well entrenched through the discourses of administrative rationalism and economic rationalism.

I argue that a fourth discourse characterised by biocentric views, which I term "green environmentalism", also profoundly affected the politics of water management in the Murray–Darling Basin. This discourse became prominent during the extended drought of the early 2000s, when efforts to protect environmental water grew more important and the green environmental movement gained influence on the basin's politics. This fourth discourse is needed to explain the policy turn that occurred during the period.

Green environmentalism elucidated how "environmental water" could be separated from "productive water", at least theoretically. Despite attempts to rectify historical wrongs by "protecting" nature, green environmentalism was grounded in problematic assumptions and came to have harmful effects. As Kay and Simmons have argued, people are a part of nature. Evidence suggests that a natural state of (pre-human) nature, as conceived by romantic assumptions embedded in green environmentalism, is impossible to identify historically, let alone restore through contemporary environmental management strategies.[15] Identifying a perfect state of nature is a subjective exercise, and aesthetic or romantic conceptions of nature do not necessarily reflect an ideal situation, in a practical sense, for animals or people.

These four discourses collectively help us understand what occurred in the Murray–Darling Basin, but none of them effectively achieved its intended goals. Instead, we saw a worsening of the crisis. Based on this research, I argue that environmental resource management should centre on the role that human societies – productive and unproductive – have in positively affecting their environments. This type of management can be achieved through knowledge of locally specific contexts and acting according to

---

15   Kay and Simmons 2002.

principles that meet the needs of local communities.[16] The most effective solutions to environmental management arise from policies that allow land managers like farmers to become empowered and to self-manage their resources. Focusing on farmers and the communities built up around them gives a fuller characterisation of a more integrated and holistic way of seeing human–environment relationships. To demonstrate the potential for bottom-up problem-solving in the Murray–Darling Basin, I piece together an emerging alternative discourse, which I term "community-centrism". By foregrounding the voices of farmers, I argue that community-centrism can help policymakers understand environmental concerns in a way that benefits human societies, nature and the long-term economic stability of communities.

Community-centrism is a response to the failings of the other four environmental discourses. In Chapters 3 and 4, I explore how these other discourses limited policy choices in response to the problems in the basin. While some farmers have learned to work within established discourses to advance their interests, these discourses still limit the bounds of acceptable discussion. My central argument in Chapter 5 is that community-centrism offers a path forward focused on social values. Building on the insights of Murray Bookchin, Elinor Ostrom and others – but grounded in the voices of the farmers I interviewed in this case – community-centrism focuses on the crucial role of community-based cooperation and engagement. This alternative discourse – focused on social outcomes – has the potential to produce complementary environmental and economic effects.

Overview

In Chapter 1, I provide a demographic and historical overview of water management in farming communities in the Murray–Darling Basin. The chapter discusses developments in the basin since colonisation and how those changes affected modern views towards water management. I describe the climate and related disasters that culminated in the crisis

---

16   Ostrom 2012.

of the Millennium Drought. Further, I explain the role of the Ramsar Convention in justifying certain political decisions in the basin in the context of an overarching administratively rational approach. The chapter also explores the challenges of creating a unified and coordinated response to water management problems. In addition, an overview of Australia's political landscape and the role of the political party the Australian Greens in defining environmental issues since the 1990s gives essential context to this discussion.

In the second chapter, I present my theoretical framework. I employ a constructivist approach to unpack the central role of ideas in discourse. While there are numerous ways to examine a discourse, this work focuses on ideas using the framework developed by Jal Mehta.[17] Mehta identified three levels of ideas – public philosophies, problem definitions and policy choices – that interact to inform policy. I focus primarily on problem definitions and policy choices to explore how they inform each other in the context of specific discourses. Chapter 2 lays out the broad contours of the five main discourses in my case study using Mehta's classification of ideas.

Chapter 3 provides an outline of the impact of the three dominant discourses in the Murray–Darling Basin: administrative rationalism, economic rationalism and democratic pragmatism. During the period of centralised and rapid modernisation, particularly during the 1950s and 1960s, little regard was shown for the social, historical and geographical context in which large environmental projects were developed. Governments dramatically altered the landscape through major irrigation projects led by the state. In this chapter, I argue that administrative rationalism has heavily affected problem definitions and policy decisions in the basin and continues to do so. Next, I examine the role of economic rationalism, which was predominant from the 1980s onwards. Governments developed market-based instruments, suggesting a more open, free-market approach to deciding where water would go. These changes sought to limit the role of government in development and trade. But adopting neoliberal policy tools fuelled a drive towards more, rather than less, government intervention. Policies related to economic rationalism had clear negative impacts on the

---

17 Mehta 2013.

community and the environment. Finally, I examine the role of democratic pragmatism in the Murray–Darling Basin. The evidence shows that these democratising processes failed to incorporate local knowledge. As a result, government interactions often reinforced centralised decision-making and increased divisive tensions. Individualism, present in farming communities and government organisations, defined the shape (and limits) of the democratic pragmatist discourse in the Murray–Darling Basin. Consequently, the discourse of democratic pragmatism is limited by the assumptions embedded within the other two central discourses discussed in this chapter.

In Chapter 4, I examine the productive effects of green environmentalism. The green movement has made significant contributions to environmentalism in Australia, but the central problem with green environmentalism is a tendency to view human societies as inherently in competition with non-human species and spaces. Government policies that separate natural environments from human environments have negatively affected the capacity of farmers to manage their environments and undermined the ability of governments to develop policies that benefit the larger ecosystems they seek to protect.

Chapter 5 presents an alternative approach to water management in the Murray–Darling Basin. I present a different view of the basin's challenges, grounded in interviews with farmers, uncovering their understanding of the issues they live with. Farmers offer a unique understanding of water management that places community interests at the top of the political agenda. Throughout this chapter, I explain how focusing on community outcomes has many positive environmental and economic consequences. Sustainable water management in the basin will depend on the government's ability to mobilise one of its most important resources: farmers. I argue that a community-centrist approach to managing water resources could lead to a greater capacity of farmers to self-manage water resources and make valuable contributions to environmental planning.

Chapter 6, titled "Policy alternatives", underscores the need for alternative perspectives, particularly those of farmers, and advocates for a more inclusive and deliberative policymaking approach. I conclude by calling for an authentic and inclusive deliberative process to redefine regional policy development and ensure the success of farmers.

The Murray–Darling Basin offers an example of how environmental problem-solving discourses inform the development of policies in ways that have consequences for both people and nature. Specifically, this case shows how dominant discourses can silence those who might offer meaningful perspectives and alternative solutions to complex and weighty environmental crises, such as drought. This conclusion has important implications as we consider policy responses to future weather extremes resulting from a changing climate. Employing Dryzek's categories of environmental problem-solving discourses and examining their interrelationships, the research shows how discourses inform the parameters of acceptable policy choices in the Murray–Darling Basin.

The analysis in the following chapters gradually reveals how policy was produced in relation to these discourses and how these discourses influence policy choices. Through a critical reflection of these discourses, we can begin to envisage an alternative future that can provide for the needs of the economy, society and environment. This work presents such an alternative view in the final chapter. I show how community-centrism offers a new way to see the synergies among the interests of nature, the economy and human communities in the Murray–Darling Basin.

1

The Murray–Darling Basin

Understanding the Murray–Darling Basin's history and the development of farm communities is critical to explaining how water management evolved to where it is today. Knowing how historical events contributed to the current hegemonic discourses is essential for unpacking and critiquing these discourses. Since colonisation, the radical transformation of the Australian landscape has made modern agriculture possible in the basin, but it has also caused irreparable environmental damage. The current situation in the basin and the proposed ecological solutions must be understood within this historical context.

The Murray–Darling Basin encompasses a wide geographical area of some 1 million square kilometres and includes New South Wales, South Australia, Queensland, Victoria and the Australian Capital Territory (ACT). The river system is vital for economic, environmental and community-based interests. The area is Australia's most important catchment, as it is home to some 2.6 million people and produces $24 billion worth of agricultural products every year. It also provides wildlife habitats and is of cultural significance to the many Indigenous peoples who live there.[1]

---

1   Australian Government Department of Agriculture, Water and the Environment 2020b; Murray–Darling Basin Authority 2020a.

13

The Murray–Darling Basin is a basin because the natural landscape collects water before it eventually flows into the two major rivers: the Murray and the Darling (Baaka). It is divided into two parts, the northern and southern basins. Water in the north runs into the Darling River, and water in the south runs into the Murray River. To the south and east of the basin are the mountains of the Great Dividing Range. Most of the rivers that flow into the basin (including the Murrumbidgee, Goulburn, Lachlan, Macquarie and Ovens) start as fast-flowing streams in these mountains. The basin is situated on flat plains not far above sea level, and the rivers in these areas tend to flow slowly. The rivers are all meandering, meaning they wind through the landscape. Because these rivers run slowly, water seeps into the land along the river, creating a fertile plain ideal for agricultural production. A large volume of water evaporates, particularly in the drier regions.[2]

Irrigated agriculture represents the bulk of farming in the Murray–Darling Basin and makes an essential contribution to the regional and Australian economies. In 2014–15, the basin accounted for 66 per cent of Australia's total irrigated land area and comprised 40 per cent of the country's irrigated agricultural businesses. These businesses include various farm enterprises: vegetable crops, tree and vine crops, pastures for grazing, hay, rice, cotton, cereals and oilseed crops. Historically, dairy farming comprised a large portion of agriculture in the basin but is now in steep decline. Average farm cash income for dairy farmers in the basin peaked in 2013–14 and began declining because of lower milk prices and higher input costs.[3] In contrast, the incomes of horticulture farms have been steadily rising.

Rice farmers comprise most of the respondents in this study. Most of Australia's rice production occurs in New South Wales in the Murray and Murrumbidgee regions in the southern Murray–Darling Basin. These areas have clay-based soils and relatively flat land suitable for rice growing. They also have well-developed irrigation infrastructure and rice storage and milling facilities.[4] Australia's annual rice production

---

2   Murray–Darling Basin Authority 2020a.
3   Australian Government Department of Agriculture, Water and the Environment 2020a.
4   SunRice 2022.

and the number of farms growing rice depend on the volume of irrigation water available to rice growers. The total area planted with rice increased gradually from the late 1980s to the early 2000s before declining significantly in 2002–03 because of drought and reduced availability of irrigation water. A return to favourable seasonal conditions in 2010–11 led to higher water allocations in 2011–12 and an increase in plantings through 2012–13. From 2013–14 to 2015–16, declining water availability resulted in falls in the total area planted with rice.[5] Rice farming has been lucrative and provided a significant portion of farm income in most cases. One of the ways rice growers manage changes in their farm operations is by adjusting the mix of agricultural enterprises each year. These farms' relatively large cropping areas (800-plus hectares) give rice growers scope to grow a mix of crops depending on markets and water availability.[6] But, as will be discussed in detail, since the Millennium Drought, rice farmers have seen a steady decline in farm income due to reduced water availability.

Another primary industry in the region is cotton. Average farm cash income for cotton growers has been growing despite dry conditions, because input costs like fuel and fertilisers have fallen. These cotton farms are mainly located in the northern basin and were not included in this study. Cotton farming has become contentious. Critics have accused these bigger cotton farms of harnessing a disproportionately large volume of water from the basin for a crop mainly for export, and argued that cotton does not support any secondary industries in the country or provide food for the nation.[7]

The tight-knit communities of the Murray–Darling Basin are supported by the farms that provide the bulk of the work to the towns. The farms I visited were almost all intergenerational but are now run by just one or two families, with most of the work done by just one or two farmers who hire contractors as necessary. Despite the vast distances between farms, farmers are neighbours to one another and work together to support each other's success.

---

5   Ashton and Van Dijk 2016.
6   Ashton and Van Dijk 2016.
7   Davies 2019.

The region's unique geographical and demographic make-up came about through more than a hundred years of settlers modifying the land to meet the expectations of European-style agriculture. The environmental costs of such a huge upending of the land are evident in the problems we witness in the basin today. The legacy of colonial settlement is inseparable from the issues in today's agricultural landscape. When European explorers first came across the area that is now called the Murray–Darling Basin, they looked out across a desert landscape and remarked that farming would never be possible.[8] When Australia became a federation in 1901, there was a widespread belief that the economic advancement of the colony would depend on its ability to harness its agricultural potential. Controlling the water of the Murrumbidgee and Murray rivers for agriculture offered an opportunity to transform these dry plains into productive agricultural lands.[9] Now, the Murray–Darling Basin is Australia's largest food bowl. This area also boasts one of the world's most elaborate and technologically advanced irrigation systems.

To understand the Murray–Darling Basin's challenges, we must first review the history of water management in the basin. Australia's treatment of the natural landscape is interwoven with its history as a colony and its treatment of Indigenous peoples. In the early days, the colonialists' attitudes towards the environment could be characterised as contemptuous. The white settlers eradicated native species, felled trees to increase pasture and stripped the earth of its natural coverings at an astounding pace.[10] Indigenous land practices, if even seen and recognised as such, were viewed as a lower form of cultivation.[11] Further, because the early colonies were nearly all run by a powerful colonial empire, the state and not individual settlers set the tone of early development. The form of governance that emerged was known as "colonial socialism", as the government was central to the growth of capital. Due to the attitudes of early settlers, the first 200 years of settlement resulted in profound changes to the natural environment.

---

8   Connell 2005.
9   Connell 2005.
10   Doyle and Kellow 1995, 2–3.
11   Pascoe 2018.

More than in other colonial nations, the changes were total and sudden.[12]

The consequences of early settlers on the landscape and the rivers of the Murray–Darling Basin have proven difficult to manage. Methods of pastoralism and dryland cropping were at the centre of the settlers' adaptation to Australia's dry climate. This adaptation brought significant technical and organisational innovation: new forms of land tenure, pasturing, disease control and stock breeding suitable to the environment.[13] During the latter half of the 19th century, new farming methods also brought technology like the stump-jump plough, the wheat stripper, mechanical harvesters and varieties of drought- and disease-resistant wheat. All these new adaptations were considered successful in adjusting farming to the climate and dry landscape of southern Australia, and technological "progress" under the direction and support of the various colonial governments was the norm. Eventually, it became clear that these techniques brought large volumes of salt previously inactive in the subsoils to the surface, including in the basin. Still, when the new Commonwealth government started enacting agricultural policies in the 20th century, it ignored these environmental problems in favour of productive large-scale agricultural development programs. Salinisation and other forms of degradation severely damaged the soil quality for future generations. More recently, it has also become clear that these techniques have severely depleted the carbon in the soil, affecting productivity in the longer term and contributing to an increasingly hostile growing environment.[14] This kind of modernisation of agriculture from the beginning of Australia's colonial history has defined the character of farming in Australia.

The Murray–Darling Basin is a closed system of groundwater basins, and the only opening is at the mouth of the Murray. Since the Murray–Darling Basin is a restricted hydrological system, the effects of human activity are evidenced quickly. Recharge rates of the groundwater system had been stable for a long time before the arrival of settlers. Even though it was a closed system, recharge rates allowed

---

12   Doyle and Kellow 1995, 3.
13   Connell 2007, 82.
14   Connell 2007, 83; Pascoe 2018.

the system to develop complex ecosystems around it. The early settlers, unfamiliar with such a unique and complex underground network of streams, did not know that if the recharge rate increased, the surface of the land and the rivers that drain the land would salinise rapidly. The extensive clearing of native vegetation and planting crops with shallow roots rapidly increased the recharge rate from rainwater to groundwater, sometimes as much as tenfold. Groundwater systems steadily rose because of these changes, which mobilised large amounts of salt that had historically been inactive in the top few metres of subsoils. Irrigation and dryland farming mobilised enormous quantities of salt accumulated in subsoils for millions of years.[15]

Despite emerging issues concerning water salinity and soil quality, the first agreement affecting the management of the Murray–Darling Basin was not concerned with the environmental conditions in the basin per se. The River Murray Waters Agreement of 1915 set out the navigation and irrigation rights of New South Wales, South Australia and Victoria.[16] This early agreement was only concerned with water quantity and did not include any provisions regarding water quality. But the irrigation and navigation planners of the late 19th century quickly realised that flow patterns along the Murray and its tributaries were much more varied than those of any other major river system anywhere in the world.[17] In 1915, a Royal Commission report noted a sevenfold difference between the highest and lowest flows of the Murrumbidgee and a tenfold difference in the Darling. The other major issue was that most of the flow came down the Murray in the winter or spring months, but plants needed to be watered in summer and autumn.[18] This meant that for European-style agriculture to ever be possible in New South Wales, a vast and complex network of dams would need to be built. Consequently, a reservoir was constructed at Albury in New South Wales and storage at Lake Victoria.[19] Two major dams were built, the Hume (completed in 1936) and the Burrinjuck (completed in 1956), as

15 Connell 2007, 86–7.
16 Doyle and Kellow 1995, 222.
17 Connell 2005.
18 Connell 2005, 85.
19 Doyle and Kellow 1995, 222.

well as a complex system of weirs, locks and canals on the Murray and the Murrumbidgee.

Massive infrastructure projects were built from the 1920s to the 1960s. Even though the Commonwealth finally acknowledged salinity problems in the 1960s, no actions were taken to mitigate the impacts of increased salinisation. The reversal of the major rivers' seasonal flow was made possible by these government-led projects meant to support large-scale European-style agriculture. A vast network of irrigation channels was built to supply various irrigation enterprises. States began to manage the water released from the storage dams into the Murray and Murrumbidgee rivers, providing water to the irrigation areas depending on the amount of available water. To this day, when water is released from the dams, it flows down the river to Berembed and Gogeldrie weirs (completed in 1910 and 1959, respectively), and from there it is diverted into supply channels that distribute the water to farms. The transition to an agricultural landscape was led by Murrumbidgee Irrigation Limited, responsible for irrigation in the region and one of Australia's largest private irrigation companies. In 1912 it was established by the NSW government after the commissioning of Burrinjuck Dam as a purpose-built scheme designed to support agriculture and provide employment opportunities. It was state-owned until 1997 and became a privatised, unlisted public company in 1999. It provides water, drainage and environmental services.[20]

A massive irrigation scheme, coupled with emerging salinity and sedimentation problems across the basin, ensured that control by state governments was essential to the system's overall operations. Federal and state governments played a crucial role in the development of agriculture, and, despite emerging environmental problems, Australia grew wealthy under the system. This transition could not come without significant environmental consequences; such consequences would make the basin entirely unrecognisable to anyone who had been living there at the beginning of the 20th century. There was a dramatic decline in native flora and fauna and the extinction of many species. For instance, a study by the University of New South Wales revealed that

---

20   Murrumbidgee Irrigation 2020.

dams and water diversions led to a 70 per cent reduction in bird species from 1983 to 2014.[21]

Despite the environmental consequences, governments continued to pour money into developing irrigation schemes and pushing for further economic modernisation of Australia's agricultural sector into the 1960s. The 1960s saw the birth of hundreds of horticultural and large-area farms around the Murray and Murrumbidgee. In that decade, a range of crops was introduced, including rice, barley, oats, legumes and grapes.[22] Wheat, olives, vegetables and cereals are also grown in some areas. These new crops were introduced alongside already established dairy and sheep farming operations. In the 1960s, a new town called Coleambally was established, and the decade also saw the growth of towns like Hay, Griffith, Narrandera and Darlington Point. More recently, cotton was introduced to the Murray–Darling Basin in 2011. Cotton plantations increased as their relative value to other crops increased over time.[23] The irrigation areas around the Murrumbidgee and the Murray have become some of the largest gravity-fed irrigation areas in the world and transformed the area into an impressively productive food bowl.

Despite growing recognition of the problems, and even a billion-dollar pledge to "fix" the Murray–Darling system, the quality of the environment continued to decline. The problems associated with water storage and management related to large infrastructure projects came to a head when the basin entered a period of drought. As early as 1991, a 1200-kilometre blue-green algae bloom formed in the Darling River. Further, in 1995 the median annual flows through the Murray's mouth were only 21–28 per cent of what they would have been in normal conditions.[24] Water was over-allocated, meaning more water

---

21 Smith 2017.

22 Rice growers, the group that most of my interviewees belong to, would eventually come to use less water than any other country in the world and produce a second cereal crop during the winter months using the moisture that is left in the soil. The largest rice processor in Australia, SunRice, is in Leeton, New South Wales. Darren De Bortoli, who I had the opportunity to interview, operates a major winery in the region and other major winemaking families also operate in the region.

23 Ashton 2019.

was permitted to be taken than was available in the rivers. The Millennium Drought (1997–2009) was the next major shock to the system. Meagre flow rates caused hyper-salinisation of the Coorong and Lower Lakes region in South Australia and the closure of the Murray's mouth. These environmental problems had flow-on effects on communities and the regional economy. Across the basin, one in every five jobs in agriculture was lost.[25] Responding to the over-allocation of water resources, the Commonwealth government's attempts at major water reform took place during a long period wherein minimal water was coming into the system. New extraction limits were based on data gathered during the drought and not before. This was problematic as it was not an accurate measurement of the average amount of water actually in the system. Nor did those extraction limits line up with what users needed from that system for various agricultural enterprises. The Millennium Drought precipitated the need to respond to the water extraction problem, but, as explained in Chapter 3, the drought made it virtually impossible to measure the effects of reform. The drought also meant it was impossible to change on-farm outcomes through reform measures.

Water management in the Murray–Darling Basin

The 1980s and 1990s were a time of great concern for international competitiveness in Australia. It was generally agreed upon in policy circles that Australia needed to become more competitive in global markets.[26] During the time, there was also a push towards rethinking government interventions in production activities. Consistent with the discourse of economic rationalism, publicly subsidised irrigation projects and policies designed to favour particular groups of farmers came under public scrutiny. Like other industries, irrigation-based agriculture became increasingly informed by free-market principles.

---

24 Wentworth Group of Concerned Scientists 2017.
25 Wentworth Group of Concerned Scientists 2017.
26 Australia became a member of the World Trade Organization in January 1995.

The drive towards government austerity hastened the free-market approach and the privatisation trend. The practice of water trading came about at this time. The stated goal was to create market-based instruments for managing water that would limit the role of government.[27]

The Millennium Drought reinforced the idea of water reform through market-based mechanisms. While there were some improvements in water conservation after a cap on diversions was implemented in 1996, this cap also presented greater economic hardship for farmers. Difficulties increased dramatically with the onslaught of the Millennium Drought. From 2001 to 2006, the Murray–Darling Basin's number of farmers and farm managers fell from 73,000 to 67,000, a decrease of 7.4 per cent.[28] At the same time, concerns over the long-term sustainability of the basin were exacerbated as drought threatened its wetlands. The national and state governments responded by examining their collective policy options. In 2002 the Murray–Darling Basin Ministerial Council released the *Living Murray* discussion paper meant to spur community consultations about whether water should be recovered from consumptive uses and put towards environmental purposes.[29] In 2003, the Council of Australian Governments (COAG) announced that member states of the ministerial council had agreed to assign $500 million over five years to reallocate 500 gigalitres (500 billion litres) of water from farms and put it towards environmental assets. Later, in 2004, state and federal governments agreed to the principle of "sustainable water use".[30] This principle led to the development of programs to secure water for environmental purposes. The most important programs are the Living Murray, which was to divert 500 gigalitres of water per year, and the New South Wales Rivers Environmental Restoration Program, which was to divert 108 gigalitres per year.[31]

27　Crase, O'Keefe and Kinoshita 2012.
28　Jiang and Grafton 2012; Murray–Darling Basin Authority 2009b, 77.
29　Crase, Dollery and Wallis 2005, 222.
30　COAG 2004.
31　Swirepik, Burns et al. 2015.

COAG reached the National Water Initiative in 2004. The parties agreed to the initiative based on the imperative to increase the productivity and efficiency of water use, to guarantee service to rural and urban communities, and to return systems to environmentally sustainable levels of extraction. Paragraph 48 of the National Water Initiative agreement states: "Water access entitlement holders are to bear the risks of any reduction or less reliable water allocation, under their water access entitlements, arising from reductions to the consumptive pool as a result of (i) seasonal or long-term changes in climate; and (ii) periodic natural events such as bushfires and drought." Further, paragraph 20 states: "The States and Territories are responsible for implementing this Agreement within their respective jurisdictions", whereas paragraph 22 states: "the Commonwealth government will assist in the implementation of this Agreement by working with the States and Territories".[32] As we can see from these passages, the agreement placed greater risk on water entitlement holders, confirmed the role of the states and territories, and gave the Commonwealth increased oversight. The government set up the National Water Commission to assist in implementing the agreement. The commission reflected the Commonwealth government's desire to assert more control over the process, mainly in response to the drought. While the National Water Initiative did not take power away from states, per se, it did hold the states accountable to the Commonwealth as an oversight body. It also solidified the commitment of the Commonwealth government to implement water reform. The emphasis was on greater efficiency in agriculture and a more comprehensive response to the river system's environmental challenges.

Most of the over-allocation at the time lay within the Murray–Darling Basin. After the more severe period of drought, which began in 1999, the Commonwealth government announced that it was better suited than the states to deal with the challenges faced by the Murray–Darling Basin. As a result, it took greater control over water management. The new role of the government was controversial on several levels. Most significantly, while the government said that future water infrastructure must meet economic standards, it offered subsidies

---

32   COAG 2004, 3, 8.

for irrigation projects in exchange for farmers selling their water allocations back to the government.

Drought, coupled with new government policies, had the effect of reducing irrigation in the basin. Irrigated land use in the Murray–Darling Basin dropped from 1,654,000 hectares to 958,000 hectares, representing a decline of 42 per cent from 2005/06 to 2007/08.[33] As such, controversy arose over the effects of water reform. But the specific effects of the reform were impossible to measure because they were accompanied by the worst drought in Australia's modern history. Despite the dramatic impact on farm communities, the Wentworth Group of Concerned Scientists recommended reducing water extractions by 65 per cent in the Murrumbidgee.[34] Two prominent members of that group, Tim Flannery and Richard Kingsford, heavily influenced the science that informed the Murray–Darling Basin Plan. The importance of their work is discussed in more detail in subsequent chapters. The Commonwealth government focused on the research and recommendations of the Wentworth Group and enforced extraction limits based on their recommendations.

The National Water Initiative provided the policy framework for institutional reform in the Murray–Darling Basin. The reorganisation of water management would be based on free-market principles, focused mainly on trade-based reform.[35] The plan promised compensation for water taken from irrigation for environmental purposes. Paragraph 49 of the initiative required:

> the risks of any reduction or less reliable water allocation under a water access entitlement, arising as a result of bona fide improvements in the knowledge of water systems' capacity to sustain particular extraction levels, are to be borne by users up to 2014.[36]

---

33 Jiang and Grafton 2012.
34 Tyson 2010.
35 COAG 2004.
36 COAG 2004, 8.

In other words, entitlement holders would have to bear the full cost of reductions meant to achieve sustainable environmental outcomes with no compensation. Paragraph 48 of the National Water Initiative stipulates that entitlement holders will not be compensated for any reductions to water meant to deal with climate change or drought, although paragraph 51 includes an agreement that governments *can* give compensation for reductions in allocations.[37] The agreement shifted responsibility for reducing allocation onto the irrigators, but there was still a drive to offer compensation to minimise disputes.

Overall, the Murray–Darling Basin Plan focused on the volume of water in the system as the cause of degradation and the key to recovery. But the consequences of altering river flows include changing sediment levels and moving salts and nutrients, which can alter the character of the main channels and wetlands. Since the beginning of the 20th century, erosion has increased significantly, filling the system with mud. When this muddy water runs into wetlands, sediment forms and blocks the sunlight, killing the plants and the native fish supported by aquatic plants.[38] If governments focus on increasing water flows but fail to address water quality adequately, they can potentially increase environmental damage. Blackwater events occur when there are elevated levels of organic carbon in the water and the oxygen levels have fallen. This can happen when there is an increase in tree litter as trees become submerged in floodwaters. In such events, the oxygen levels drop so low that fish populations suffocate and die. Thermal pollution, when water is at the wrong temperature, can have serious negative consequences, like preventing fish from being able to breed.[39] The large number of blackwater events in the basin since the plan has been put into effect has elicited criticism from farmers that the government is paying too much attention to water quantity and not enough attention to water quality.

A century of re-engineering the river system has created fundamental changes that have dramatically altered the natural ecology and damaged native fish populations. Even when recovered water is

---

37 Connell 2007; COAG 2004.
38 Russell 2017.
39 Russell 2017.

of good quality and the benefits of increasing volumes can be seen, governments also must weigh up if a higher volume of water is changing the temperature of the water. Further, increasing the volume of water does not address the numerous hazards and obstacles that native fish must face in an irrigation system that supports farms, not fish. Other mitigating measures include restocking threatened fish species, eradicating carp and other non-native species that destroy the ecology of the river, overhauling irrigation infrastructure such as screens on irrigation pumps or overshot weirs so that they can accommodate fish populations, and improving the habitats around floodplains (on farms, and in state and national parks).

The government has maintained a hardline approach to its target of recovering 2,750 gigalitres of water for environmental purposes.[40] The Murray–Darling Basin Authority is responsible for coordinating all the monitoring that happens, both from an ecological and socio-economic perspective. The Commonwealth Environmental Water Holder, part of the Department of Climate Change, Energy, the Environment and Water, does its own environmental water monitoring to look at the effectiveness of its decisions. There is no compulsory acquisition, but the Commonwealth buys water from willing sellers, which is politically sensitive. These problems will be discussed in detail in Chapter 3.

There are significant difficulties in terms of managing water resources in the basin. In addressing these challenges, governments and researchers must first understand the history that shaped the policy landscape. Historically embedded assumptions and habits contribute to ways of understanding and communicating these problems. Considering this, as explained in the next chapter, discursive analysis provides an ideal tool for understanding these problems.

The Murray–Darling Basin is economically, environmentally and culturally of crucial significance for Australia and has become the "breadbasket" the country. But the history of the Murray–Darling Basin is defined by colonisation in Australia and the imposition of European conceptions of farming on the land. The environmental consequences of this historical legacy have been profound, with salinisation and overuse of precious water reserves defining the course of water

---

40   Murray–Darling Basin Authority 2019c.

management over time. Further, we can see how a history of administrative rationalism affected state modes of interference in environmental water management as the tide shifted towards environmental responsibility. The challenges of water management in the early 21st century coalesced with the drought and produced the worst crisis in water management in Australian history. Notwithstanding a great deal of attention and attempts to manage the Murray–Darling Basin over the last century, there remain deep problems associated with competition over this resource and claims of mismanagement on all sides. How can we make sense of what has happened here, particularly regarding the failures in policy design and implementation? In what follows, I unpack how discourses helped construct the crisis and their impact on policy choices and outcomes. But first, the next chapter explains the theoretical approach of this research.

2
Theoretical approach

Discourses play an essential role in water management decisions in Murray–Darling Basin farming communities. Discourse is defined herein as a set of interrelated texts and practices of their production, dissemination and reproduction that bring an object into being. Social reality is produced through discourse, and social interactions cannot be understood without reference to the discourses that give those interactions meaning.[1] At the same time, these interactions, practices and routines reproduce and normalise discourse. This book explores how discourses affect the policy preferences of farmers and policymakers and how discourses shape acceptable policy choices. Further, the work examines how farmers resist discourses and what alternative perspectives, knowledge and policy options are excluded by these discourses.[2]

A central precondition for reciprocal dialogue between the state and communities is the openness to alternative ideas, perspectives and

---

1   Phillips and Hardy 2002, 3.
2   I focus primarily on farmers as the vehicle for change within the Murray–Darling Basin as they control much of the land there that is under cultivation. Indigenous communities also have much to contribute in terms of sustainable resource practices, but they do not exert as much influence in this particular region. Several attempts were made to locate Indigenous farmers, but with no success.

knowledge.[3] Expert planners and farming communities have distinct ways of understanding the world, as evidenced through discursive practices like the language, rituals and symbols each group uses. Discourses define the parameters of what is considered acceptable and desirable. Uncovering the assumptions embedded in discourse is essential to understand how interests are defined and how they can be redefined.

In this chapter, first, discourse analysis is defined within the context of the broader literature. Discourse has a role in constructing, communicating, challenging and institutionalising certain ideas over others. Second, Jal Mehta's framework of ideas is used to understand how discourses manifest in practice and affect political decision-making.[4] Third, I position my work within a broader literature on water governance theory and explain my contribution to this literature. Finally, I outline the five discourses central to the analysis undertaken in this work.

Discourse analysis reveals how power is embedded within and reproduced through the ways we communicate with one another. French sociologist Michel Foucault provided the most detailed account of the power of discourse. He argued that discourse has both productive and disciplinary effects. Discourse is productive when it leads to certain consequences or ends. Discourses are productive because they shape possibilities, ideas, beliefs, values, identities, interactions with others, and our behaviour. A discourse also encompasses a claim to truth that has disciplinary power.[5] Discourses can be disciplinary by limiting acceptable policy options. Similarly, Hajer argued that discourses can make it difficult to raise certain questions. According to Hajer, only certain people are allowed to participate in a discourse because it contains internal disciplines through which the prevailing order is maintained.[6] Discourse includes written and spoken words, how we depict the world through symbols and pictures, and how we engage with it through practices and routines. Everyone functions within

---

3   Dryzek 2013; Hajer and Versteeg 2005; Litfin 1994.
4   Mehta 2013.
5   Foucault 1980.
6   Hajer 1997, 49.

certain discourses and associated practices, but we often take for granted the discourses we operate within. As Foucault explained, even though discourses have productive and disciplinary power, this does not mean they cannot be challenged or that the experts who define a discourse will always control it. Resistance can emerge from within a discourse to question its own norms.[7]

In this work, I unpack the discourses at play in the Murray–Darling Basin to reveal their productive and disciplinary effects. I also look at how these discourses are challenged through various forms of resistance. Fleming and Vanclay have written that resistance in discourses is a site for agency and transformation.[8] This understanding is influenced by Foucault, who argued that there is typically a dominant discourse that can only be challenged by competing perceptions within it, which he termed "resistance". There are, therefore, productive powers inherent within a discourse (as opposed to discourse's disciplinary power):

> Discourses are not once and for all subservient to power or raised up against it … We must make allowances for the complex and unstable process whereby a discourse can be both an instrument and an effect of power, but also a hindrance, a stumbling point of resistance and a starting point for an opposing strategy. Discourse transmits and produces power; it reinforces it, but also undermines and exposes it, renders it fragile and makes it possible to thwart.[9]

In this research, I draw on Foucault's perception of resistance to show how farmers work within the contours of defined discourses to find ways to influence policy outcomes to achieve their ends.

Richardson, Sherman and Gismondi used the concept of resistance in their discursive analysis of the environmental impact assessment hearings for the Alberta-Pacific bleached kraft pulp mill held in Alberta and the Northwest Territories in 1989–90. They examined how parties

---

7   Foucault 1978.
8   Fleming and Vanclay 2009.
9   Foucault 1978, 100–1.

opposed to the project, including farmers, trappers, Indigenous groups and others, used the dominant discourse as a starting point for presenting an alternative viewpoint. Opponents of the mill pointed to the common use of the phrase *sustainable development* as an example of how language is used to mask the objective of rapid economic growth and make exploitation of the natural environment more palatable.[10] Opposing parties were able to present an alternative reading of sustainable development that aligned with its core assumptions but was more amenable to their own goals. Like Richardson, Sherman and Gismondi's work, this research gives examples of farmers seeking to contest, subvert, rework and supplant dominant discourses. This is done – in part – by challenging, using and reinterpreting commonly accepted phrases and terms.

It is important to point out some distinctions between the works of Foucault and Dryzek – given how much this work draws on the latter – and where this research is positioned. Foucault's writings focus on the power of a dominant discourse and how that discourse can be challenged from within. Dryzek, in contrast to Foucault, argued that competing discourses act simultaneously, with some seeking to undermine or supplant others. Foucauldians are committed to the idea that people are subject to the discourse within which they move and cannot step back and make choices across different discourses. Dryzek disagreed: "discourses are powerful, but they are not impenetrable".[11] While Foucault argued that one single discourse is dominant to a particular field or issue, Dryzek contends that hegemonic discourses can erode and be supplanted by other discourses that may have emerged in relative isolation from the dominant discourse. Further, competing discourses are often quite distinct, potentially offering differing views that may constitute an alternative hegemonic discourse.[12] In Foucault's theory, discourse is singular and dominant, with both productive and disciplinary effects. In Dryzek's theory, multiple discourses exist simultaneously and have productive and

---

10   Richardson, Sherman and Gismondi 1993, 51.
11   Dryzek 2013, 22.
12   Dryzek 2013, 22.

disciplinary effects. This work builds primarily on Dryzek's theoretical position.

This case reveals several competing and conflicting discourses. Dryzek conceptualised such conflicts as an opportunity for participants – as well as the analysts studying them – to see that there are different ways of addressing a specific challenge. In other words, conflicts between existing discourses open the possibility for other discourses and, thus, new policy solutions. But such conflicts can result in discourses losing steam and becoming realigned with dominant frames of reference. For example, the mass demonstrations and protests of the Occupy movement, which began in the United States but eventually came to contest economic and social inequality all over the world (2011–12), changed the discourse within society ("the 99 per cent vs. the 1 per cent") and challenged the dominant economic order. However, over time the movement's demands were refined to fit within the economic order. Its calls for action, such as "take money out of politics" or "increase taxes for the rich", represented compromises designed to make its message more palatable within the established political and economic order. These compromises came to form part of the Democratic Party's political platform in the United States.[13] For the Occupy movement, the seeds of a new discourse arose out of direct conflict with the status quo. Eventually, it came to reflect a compromise within the established economic discourse.

The co-constitutive relationship between discourses and practices is explored in this work. Several other authors, such as James Scott and Tania Murray Li, have taken a similar approach. In his book *Seeing Like a State*, Scott argued that discourses exhibit agency. Scott offered the notion of transcripts, both hidden and public. These are established ways of behaving and speaking that fit particular actors in specific contexts. This idea can be linked to Foucault's notion of resistance within discourse. Actors resist public transcripts by using prescribed roles and language to fight the abuse of power.[14] In the case of the Murray–Darling Basin, farmers have made use of such transcripts to resist government actions. But it is essential to recognise that both

---

13   Levitin 2015.
14   Scott 1999; 1985, 137.

dominant and weak parties are often caught within the same web of socialised roles and behaviour.[15] These roles are frequently expressed without any conscious intent. In this sense, power structures can be subconscious and internalised through transcripts rather than deliberate, intentional and calculated acts of domination.[16] In the Murray–Darling Basin, this is often the case. Farmers, consciously and unconsciously, think and communicate within the prescribed norms of a dominant discourse even as they resist power structures. In lay terms, farmers operate within the confines of the status quo, even when resisting the status quo. For example, farmers use transcripts from the discourse of economic rationalism to challenge dominant policy prescriptions and present their priorities in terms that are, arguably, even more economically rational than government policies.

One of the critical questions this research addresses is how local farmers have challenged the dominant discourses in the Murray–Darling Basin and put forward their own knowledge and policy solutions. To address this requires thinking critically about the relationship between local farmer knowledge and the state. From a cynical point of view, eliminating local (in this case, farmer) knowledge can be seen as a prerequisite for asserting state power. A more optimistic view, as expressed herein, is that there can be a reciprocal relationship between local and state knowledge. Scott described the former view in *Seeing Like a State*. Scott observed that governments act in ways that increase their power, whether intentionally or not. Given the nature of power, he argued that states devise improvement schemes based on the kinds of information that will allow states to intervene and assert their authority. For Scott, due to the high modernist nature of many states, the elimination of local knowledge becomes a precondition for administrative interventions, taxation, worker disciplines and profit.[17] While Scott's arguments are compelling in that

---

15   Scott 1992.
16   Scott 1992.
17   Scott 1999. High modernism is characterised by an unfaltering confidence in science and technology as means to reorder the social and natural world; the underlying assumptions of high modernism are closely aligned with administrative rationalism and discussed further below.

they draw attention to how governments and elites try to suppress local knowledge to push their agendas, his approach does not pay enough attention to instances in which local actors *can* influence governments. At any given time, numerous actors influence the government's agenda.

Tania Murray Li, in "Beyond 'The State' and Failed Schemes", takes the latter, more optimistic approach to understanding the relationship between states and local people.[18] She argued that eliminating local knowledge is not a *necessary* consequence of state power. To support her argument, she gave several examples of how experts have used local knowledge. Li's theoretical perspective highlights the role of non-state actors that attempt to influence government: particularly social reformers, scientists and non-governmental agencies. Like Scott, Li asked why specific schemes to improve the human condition have failed, but she reframed the question to ask: what could these schemes potentially do? For Li, there can be processes that allow for a productive relationship between planners and those affected by improvement schemes. Identifying these processes requires a closer examination of how relationships are socially constructed between these actors. This research, therefore, carefully examines the social relationships between farmers and policymakers in the Murray–Darling Basin and demonstrates Li's conclusion that discourses are shaped and transformed by the social relationships of actors.

Foucault argued that one cannot only conceptualise power as a manifestation of material relations, but that knowledge is its own form of power with distinct origins and effects. One environmental and political scientist who works with Foucault's ideas to analyse productive and disciplinary power is Karen Litfin. I adopt Litfin's view that discourses determine what can and cannot be thought, thereby defining the range of policy options. She wrote: "discourses entail, but are not reducible to, interpretations. Rather, they are broader sets of linguistic practices embedded in networks of social relations and tied to narratives about the construction of the world".[19] Further, as Foucault discussed, discourses empower certain actors and exclude others, although they also offer a site of resistance that can provoke new

---

18   Li 2005.
19   Litfin 1994, 252–3.

discourses. Under conditions of scientific uncertainty, particularly in the case of environmental crisis, actors of all types can reframe and interpret information in ways that eventually change the discourse. Ecological problems, therefore, are "discursive phenomena that can be studied as struggles against contested knowledge claims, which become incorporated into divergent narratives about risk and responsibility".[20] In Litfin's study of how the precautionary principle came to dominate debates about how to respond to the atmospheric ozone "hole" in the late 1980s and early 1990s, she explained that, while scientific uncertainty justified a cautious approach, a shift in the meaning of the word "caution" itself came to be significant. Reframing "caution" to highlight the potential environmental consequences of an expanding hole in the ozone layer suddenly made environmental vulnerability more acute than industrial vulnerability.[21] Litfin believed that the shift did not emerge from scientific consensus alone but in combination with specific discursive strategies. The role of agents in influencing the discourse is therefore critical in terms of policy definitions and solutions.

Understanding the assumptions embedded in environmental discourses and the dynamics within and between discourses – and the actors who use them – is vital to understanding policy prescriptions and outcomes. Discourse analysis helps us see the epistemological and ontological assumptions about the world embedded in policies and day-to-day life. Acknowledging and understanding these underlying assumptions will ultimately be essential for addressing problems more effectively through policy measures.

While discourse analysis provides a productive method for understanding the problems of the Murray–Darling Basin, there are many possible approaches to conducting discourse analysis. One such point of departure is Jal Mehta's framework, which looks at discourses in terms of how they shape policy at the level of ideas. This work uses elements of Mehta's theoretical approach to understanding how specific ideas function in the context of the Murray–Darling Basin. While discourses embody more than just ideas – and the relationship

---

20   Litfin 1994, 254.
21   Litfin 1994, 260.

between ideas and practices within a discourse is co-constitutive – focusing on the power of ideas is valuable in and of itself. Mehta's framework helps to unpack key ideas within discourses and to understand how these translate into specific policy decisions. Mehta considered ideas at three levels of generality: policy solutions; problem definitions; and public philosophies or zeitgeists. A policy solution is the narrowest form of an idea. Problems and solutions are not pre-established. The way a problem is defined has significant implications for the types of policy solutions that become desirable. Much political contestation therefore occurs at the level of problem definition. Public philosophies and zeitgeists represent the third, higher-order ideas. These are ideas that cut across substantive areas. For example, public philosophy includes ideas about the role of the government or public policy in the context of broad assumptions about the market or society.[22]

What Mehta termed public philosophy is akin to what Dryzek identified as the broader assumptions embedded in environmental problem-solving discourses. Mehta described public philosophies as influencing the kinds of problem definitions that emerge. In turn, problem definitions narrow the range of possible solutions and the type of knowledge(s) considered relevant for solving a problem. Nevertheless, certain forms of new knowledge can make their way into dominant discourses if that knowledge is constructed in a way that is consistent with the prevailing public philosophies. Public philosophies, or what Dryzek called the overarching problem-solving discourse, can and do change over time, but it is noteworthy that they are highly resistant to change. The ontological and epistemological assumptions embedded in an environmental discourse, as Dryzek discussed them, are essential elements of public philosophies.[23]

Of Mehta's three levels of ideas, problem definitions are the primary focus of analysis in this research. As Mehta explained, political arguments are fought mainly at the level of problem definition.[24] Problem definitions tend to combine both normative and empirical

---

22　Mehta 2013.
23　Dryzek 2013.
24　Mehta 2013.

claims, with these two elements being mutually reinforcing. The way that problems are framed has significant implications for the kinds of policy solutions that emerge; these are productive effects of discourse. On the disciplinary side, problem definitions can limit or narrow the scope of desirable policy solutions while dismissing what might be considered illogical or misguided. Focusing our attention on discourses at the level of problem definition can help reveal underlying public philosophies while at the same time explaining chosen policy solutions and their effects. In other words, focusing on problem definitions allows us to uncover higher-order ideas and assumptions (the broader public philosophies) and the disciplinary effects of discourse on lower-order ideas that take the form of policy solutions.

As Mehta explained, the narrowest form of ideas is policy solutions. At this level, we can look at the actions and responses that emerge from contests over problem definition. We also need to look at the behaviours and ways of communicating that appear along with a given discourse to reinforce its central assumptions or challenge them. When considering this level of ideas, James Scott's notion of "transcripts", both hidden and public, is helpful. As discussed above, transcripts are established ways of behaving and speaking by particular actors in specific contexts.[25] Scott developed this idea in a way that links to Foucault's notion of resistance within discourse. Actors resist "public transcripts" by using prescribed roles and language to resist the abuse of power.[26] Notably, Scott argued that the dominant and the weak are often caught within the same web of socialised roles and behaviour. These roles are frequently expressed without any conscious intent. In this sense, Scott had a cultural and psychological view of power structures as subconscious and internalised through transcripts rather than deliberate and calculated acts of domination.[27] This view aligns with Dryzek's conception of discourse as having a reflexive and shifting nature. People, consciously and unconsciously, think and communicate within prescribed norms even when resisting dominant power structures. The theoretical concept of transcripts is used in this

---

25 Scott 1992.
26 Scott 1985, 137.
27 Scott 1992.

analysis of Murray–Darling Basin politics. For example, farmers often use transcripts from the discourse of economic rationalism to challenge dominant policy prescriptions and present their priorities in, arguably, economically rationalist terms. These transcripts from the government and farmers form an important aspect of this analysis.

Mehta's conception of ideas, particularly "problem definition", is used herein to analyse Murray–Darling Basin politics. A problem definition is when we seek to define an approach to a given problem. Problem definition operates in the background, entering discussions as arguments are made for or against a policy.[28] New problem definitions allow debates to evolve and change, and they also affect the kinds of practices that come to be accepted. Practices also influence which ideas are accepted and which are not. Mehta wrote that problem definition is a contested process among players with differing amounts of power and persuasiveness. He identified six factors that will determine whether a given problem definition succeeds. First, the powers and resources of the claimants: power is not limited to the resources of the actors; it is also about the ability of actors to frame issues in ways that give them power. Second, problem definitions are successful based on how claimants portray the issues (framing). There is a wide range of strategies available to actors to successfully define an issue, including compelling storytelling, shifting the burden of proof, using accepted metaphors and invoking symbols. Third, the venue or context in which the problem is heard affects the success of problem definitions: shifting venues can effectively garner support from actors. Fourth, establishing authority over a problem definition is a key battleground for lending legitimacy to any problem definition. Fifth, whether there is a policy solution for a given problem and definitions determines the success of problem definitions: successful problem definitions are generally accompanied by strong and viable policy proposals. Finally, the fit between the problem definition and the broader environment affects the success of the problem definition. Whether a policy definition resonates with the views held by the public or media is a key determinant of whether a problem definition will be successful.[29] These

---

28   Mehta 2013.
29   Mehta 2013.

six factors appear in various forms throughout this study. For instance, farmers try to frame problems in ways that give them the moral high ground. Both farmers and government officials are also careful to couch their issues in language that has the most impact in garnering public support. This strategy is evidenced by the metaphors and rhetoric they use. Further, both sides seek to establish authority over how problems are defined by representing themselves as experts.

What typically occurs within policy development circles is that actors who offer problem definitions that conform easily to established sets of practices and policy solutions tend to have more influence. Foucault referred to this as the productive effects of discourse.[30] At the same time, dominant problem definitions discipline the range of policy solutions by narrowing the possibilities considered acceptable. Actors who offer solutions that do not reflect the prevailing problem definitions have difficulty advancing those solutions. Dominant problem definitions can, therefore, discipline the search for solutions and the evolution of what may be acceptable. Problem definitions can also act as sites of resistance, as actors try to change problem definitions in ways that will be accepted and thereby change policies and practices. Under Dryzek's framework of competing discourses, shifts in problem definitions can be influenced by shifts towards or away from certain discourses. For instance, the rise of green discourse (with its distinct public philosophy) arguably contributed to significant shifts in how policymakers defined problems in the Murray–Darling Basin.

Dominant problem definitions also determine who has the perceived legitimacy to speak. For example, if water is defined as a tradeable asset, economists may have more legitimacy to talk about water than environmentalists. But, consistent with Foucault's and Scott's understanding of the politics of resistance, when marginalised voices start working within dominant discourses to redefine how problems are expressed and addressed, they *can* also have an impact (though this can be an uphill battle). In sum, a dominant problem definition is a form of power that can reinforce the status quo, but contestation over problem definitions can also create opportunities for significant institutional change. New problem definitions may allow for

---

30   Foucault 2002.

policy options that do not necessarily conform to the dominant policy paradigms, thereby contributing to a shift in public philosophy.[31]

Table 2.1 depicts my theoretical framework's main elements, highlighting how Mehta's categories of ideas connect with Dryzek's elements of a discourse. Reading across the table shows the alignment between Mehta's categories and Dryzek's elements of discourse analysis and the questions raised at each level that we can bring to the study of specific cases. Reading down the table, we find each of Mehta's three levels of ideas. The third level, that of policy solutions, also incorporates Scott's concept of transcripts and Foucault's notion of resistance as "elements of discourse" to explain the discursive struggles that occur at the level of developing and implementing policy solutions. This table also provides examples with reference to Dryzek's three problem-solving discourses: administrative rationalism, economic rationalism and democratic pragmatism.

---

31 Hall 1993.

Table 2.1 Categories of ideas and elements of discourse

Categories of ideas (Mehta)	Elements of discourse	Analytical questions raised
Public philosophy Includes values, beliefs about the economy and the role of government in society. Includes normative statements about the scope and nature of government in society. Is closely aligned with what Dryzek refers to as ontological and epistemological assumptions.	Basic entities recognised or constructed *and* assumptions about natural relationships (Dryzek) Example: Under administrative rationalism, liberal capitalism is a basic entity. Basic assumptions include that nature is subordinate to human problem-solving and that experts and managers should control environmental decision-making. Example: Under economic rationalism, markets and private property are key basic entities. Basic assumptions include that competition is good and that the environment is best regulated through free market principles.	What are the basic entities constructed and re(affirmed) through this discourse? How is the role of the state defined in this discourse? What is the relationship between individuals and the markets in this discourse? What is the nature of competition and cooperation in this discourse? How is the relationship between nature and people defined in this discourse? How is "environment" understood in this discourse?
Problem definitions A problem definition is a particular way of interpreting an issue. A problem's framing has significant implications for how proposed policy solutions are constructed and perceived.	Assumptions about agents and their motives, *and* key metaphors and other rhetorical devices (Dryzek) Problem definitions provide an explanation of the nature of a problem, including its severity, impacts, relevance and causes.	How do problem definitions understand the ways that actors are motivated? What common metaphors and rhetorical devices are used to characterise the problem?

42

Problem definitions act as a site of resistance within a given discourse (Foucault). Example: The water crisis may be defined as being caused by poor farming practices, or by the selfish interests of individual farmers, or by climate change (with each definition relating to competing normative, ontological and epistemological assumptions identified above as well as the contests over policy solutions identified below).	Example: Under democratic pragmatism, it is assumed that actors are motivated both by material self-interest and by multiple conceptions of public interest.	Example: Under economic rationalism, it is assumed that actors are motivated by self-interest and that some government officials may be motivated by public interest. Rhetorical devices often include "market freedoms" and a rejection of "command and control".
Policy solutions The proposed solutions deemed relevant to specific problem definitions. The specific policy tools and practices of government, which in turn shape the perceptions and behaviours and practices of a wide range of actors. Example: Under administrative rationalism, policy solutions may include strict government regulations to regulate behaviours, such as forced reductions in consumptive use or financial penalties for over-extraction.	Transcripts (Scott) *and* actions/ responses that emerge in support of or resistance/opposition to the policy solutions (Foucault) Example: Under economic rationalism, if policy solutions focus on market-based incentives, we may then see farmers use terms associated with economic rationalism to challenge the actions of the government.	What are the words, phrases and actions that frequently appear in this discourse to (re)produce its central assumptions? What are the specific orientations and policy solutions proposed? What alternative solutions or ideas are marginalised by the policy solution, and which actors are excluded/included from it? What resistance is generated within the discourse associated with this policy solution, and what effects does this resistance have?

Political scientists working in the realm of ideas – known as constructivists – generally take a defensive posture, trying to establish that ideas matter in a discipline that privileges neo-Marxist, structuralist or rational-choice models of explanation.[32] Constructivists argue that material forces alone are insufficient in explaining how people act. Mehta explained that we must move beyond whether ideas matter and ask *how* they matter. Mehta wrote:

> ideas, broadly defined, are central to questions about agenda setting, social movements, revolutions, diffusion, policy choice, the conceptual categories that underlie politics, path dependency and path-shaping change institution building, institutional stability, institutional change, voter identity formation, interest group formation, and political coalition building.[33]

Ideas matter in terms of their effects on the policies that emerged in the Murray–Darling Basin. The assumptions embedded within each discourse are grounded in a specific public philosophy, and each public philosophy, in turn, relates to problem definitions and policy solutions. While there are other aspects of discourse apart from ideas, focusing on these three levels of ideas has significant explanatory power. The case demonstrates that change may occur at the level of public philosophies – particularly when a crisis occurs – allowing new voices and perspectives to gain power. For example, in the case of an environmental crisis like drought, an epistemological crisis can also occur, enabling certain public philosophies to gain more currency than others.[34] The term "drought" is a discursive way of framing low rainfall over an extended period in a way that stimulates a new problem definition. Problem definitions, in turn, tend to dictate the possibilities and limits of acceptable policy solutions. Important changes happen at the level of problem definitions. Water, for instance, can be defined as an asset, a commodity, a resource or an element of the natural

---

32   Mehta 2013.
33   Mehta 2013, 291.
34   Whether or not a shift in public philosophy actually occurred in this case is something I explore in Chapter 4.

world. Each definition has consequences for how we understand water and what can be done with it. The way we define problems is rooted in language, norms and values. Change can also stem from policy solutions by shaping actors' practices, roles and transcripts. Policy solutions, in turn, can help shift problem definitions and, ultimately, public philosophies (though the latter are notoriously persistent).

While distinctions among the discourses emerged in the case of the Murray–Darling Basin, it is important to acknowledge that these discourses sometimes overlap, and one discourse can supplant another. In this work, discourse analysis is focused primarily on the level of problem definition but also looks at how a crisis like a drought contributed to shifting public philosophies. While discourse analysis includes more than just ideas, Mehta's conception of ideas has significant explanatory power in the case of the Murray–Darling Basin and thus is the primary focus of analysis.

Water governance theory

Governance in the Murray–Darling Basin must be understood in the context of the broader literature on water governance. Water governance refers to the political, social, economic and administrative systems in place to develop and manage water resources and deliver water services at different levels of society. The notion of governance for water includes the ability to design public policies and institutional frameworks generally accepted by society. Water policy has the critical goal of sustainable development of water resources, and key stakeholders must be involved in the process.[35] Water governance theory is about how water is governed *and* about prescribing policy approaches based on what can be learned from a diversity of contemporary cases. The following provides a broad overview of the current water governance literature, explaining key concepts like polycentric and collaborative governance, and how these concepts relate to the case of the Murray–Darling Basin.

---

35   Rogers, Hall and Wouters 2008.

There is an overall agreement in the water governance literature that polycentric governance structures are desirable.[36] This literature argues that shared structures of power are more effective than other models, including market-based allocation, as well as both centralised and decentralised systems.[37] Polycentric governance systems are distinct from each of these other three models. In markets, individual citizen-consumers are responsible for providing goods and services. In centralised systems, democratically elected governments make decisions over the supply of public goods like water on behalf of citizens. In decentralised systems, authority is allocated to sub-national (for example, regional and local) governments. In polycentric governance systems, different authorities – central and local governments, agencies, self-governed user groups, firms, or other hybrid organisations – participate in markets and plan horizontally and vertically across geographic scales to co-produce public goods. Polycentric systems have multiple centres of power, and these generally are non-hierarchal.[38] Pahl-Wostl and colleagues argued that an essential condition for improving performance is striving for more polycentric structures since polycentricity allows river basins, regions and countries to find a governance structure uniquely suited to their circumstances rather than following narrowly prescribed courses of action.[39]

The earliest works on polycentric governance concern local communities' self-governing capacity. The assumption underlying this literature is that, since local communities face unique sets of problems, their local knowledge means they are in a better position to address problems. Polycentric systems are seen to be more resilient because issues with varying geographical scopes can be managed at different levels. If one form of governance fails, another may be more successful. Further, the large number of units means there are more opportunities

---

36 Garrick, Heikkila and Villamayor-Tomas 2018; Huitema, Mostert et al. 2009; Pahl-Wostl, Arthington et al. 2013.
37 Pahl-Wostl, Arthington et al. 2013.
38 Garrick, Heikkila and Villamayor-Tomas 2018; Huitema, Mostert et al. 2009.
39 Pahl-Wostl, Arthington et al. 2013.

for experimentation with new approaches, and the units can learn from one another.[40]

It is important to note that water governance in the Murray–Darling Basin can be thought of as polycentric in some ways. In the case of the Murray–Darling Basin, local entities like councils, regional groups like catchment authorities, and provincial governing entities have historically aligned decision-making with national policy.[41] Governance has depended on a broad consensus about rules, policies and values because no one party was in charge.[42] These alignments suggest that it could be useful to examine the case of the Murray–Darling Basin through the lens of polycentric governance.

The literature on environmental governance describes different approaches to polycentric governance. For example, the "classical modernist" approach to polycentric institutional design is organised by jurisdictions within hierarchical government levels (national, regional and local) without overlaps in tasks.[43] Jason Alexandra drew on this understanding to unpack "good governance" structures in the Murray–Darling Basin. Alexandra argued for several institutional features that historically endowed capacities for trans-boundary governance in the Murray–Darling Basin. First, shared waters meant shared infrastructure that required shared decision-making frameworks. Second, formal intergovernmental agreements provided the constitutional frameworks that enabled further agreements, particularly regarding salinity. Third, tensions between different governments required negotiations in which pragmatic solutions generally prevailed. Fourth, attempts to organise across geographical boundaries and technical, scientific and policy domains allowed information to be shared between agencies, experts, water users and the public. Finally, diverse partnerships with educational and research agencies, community and industry groups established a collective sense of responsibility for co-managing rivers.[44] Despite the increased

---

40   Huitema, Mostert et al. 2009.
41   Wyborn, van Kerkhoff et al. 2023.
42   Abel, Wise et al. 2016.
43   Hajer 2003.
44   Alexandra 2019.

centralisation of governing authority for water resources under the Murray–Darling Basin Authority since the Millennium Drought, the classical modernist approach to institutional design, where an exclusivity of jurisdictions is prominent, is still prevalent in the Murray–Darling Basin.

The classical modernist design is now increasingly regarded as impracticable and ineffective.[45] Water governance strongly depends on institutional arrangements. As Alexandra wrote: "the plethora of recent inquiries into the adequacy and integrity of governance arrangements in the Murray–Darling Basin indicates a crisis of trust, legitimacy and public confidence – in short, a loss of authority".[46] Current governments are losing the authority and legitimacy needed to govern the basin despite efforts at more collaborative governance. Inquiries have revealed serious concerns about integrity, poor administration and failures of compliance and enforcement regimes.[47] Further, recent reforms in the Murray–Darling Basin have marginalised actors outside federal, state and territorial governments. Political conflicts between central governing bodies have impeded progress towards water reform. The flexibility and adaptability that characterise polycentric governance seem to be lacking in this case.[48]

This classical modernist approach is closely aligned with Dryzek's discourse of administrative rationalism, which focuses on the role of government institutions and bureaucracies in orchestrating effective water governance.[49] In the last 20 years or so, worldwide, the state's historical role of directing society has been contested by unified local networks (civil society, private sector) and global networks (international organisations and non-government organisations). The state is increasingly seen as part of the problem rather than the solution. Further, states no longer believe they can solve societal problems acting alone, particularly socio-environmental ones. There is also increased awareness that markets alone will not be able to address social and

---

45 Huitema, Mostert et al. 2009.
46 Alexandra 2019, 99.
47 Alexandra 2019.
48 Wyborn, von Kerkhoff et al. 2023.
49 Dryzek 2013.

environmental problems. Consequently, both hierarchical governance and market-led models are weakened.[50] As is explained in my discussion of administrative rationalism and economic rationalism in this chapter, both state and market-led approaches are being challenged by more bottom-up approaches. In response to these challenges, governance models – especially ones we can define as polycentric – have moved towards a more collaborative approach.

Ansell and Gash defined collaborative governance as "a governing arrangement where one or more public agencies directly engage non-state stakeholders in a collective decision-making process that is formal, consensus-oriented and deliberative and that aims to make or implement public policy or manage public programs or assets".[51] This definition tries to emphasise six criteria: public agencies or institutions initiate the forum for deliberation; participants in the forum include non-state actors; participants engage directly in decision-making and are not merely "consulted"; the forum is formally organised and meets collectively; the forum aims to make decisions by consensus; and the focus of collaboration is on public policy or public management.

As we can see, one of the main attributes of collaborative governance is public participation. Public participation is thought to improve decision-making by expanding the decision-making process to more people. Participation can enhance public understanding of the issues, make better use of information from the public, make decision-making more transparent, and encourage governments to respond more thoughtfully to input that is received.[52] Overall, participation improves transparency and contributes to more democratic decision-making. But public participation is only feasible if there is a willingness and capacity of stakeholders to participate and of policymakers to organise participation efforts. Interest groups must have the resources and level of organisation to participate effectively. Further, policymakers are not always willing or able to invite public participation, especially if the management culture is technocratic.[53] Collaborative strategies face many obstacles: influential

---

50   Rogers, Hall and Wouters 2008.
51   Ansell and Gash 2008, 544.
52   Huitema, Mostert et al. 2009.
53   Huitema, Mostert et al. 2009.

stakeholders manipulate the process; public agencies lack real commitment to collaboration; and distrust becomes a barrier to good faith negotiation. Nonetheless, collaborative strategies have sometimes meant that adversaries have learned to engage in productive discussions, public managers have developed better relationships with stakeholders, and alternative forms of collective learning and problem-solving have emerged.[54] This literature on collaborative governance is closely aligned with assumptions embedded in the discourse of democratic pragmatism introduced by Dryzek, wherein there is a belief that participation is key to effective democratic decision-making. The literature on democratic pragmatism underscores these positive benefits of enhanced efforts at collaborative governance. But governments have been reluctant to move towards a more collaborative approach to water management. In cases examined by Huitema and colleagues, the researchers found that governments lacked experience with multi-party processes, relied heavily on the expertise of experts, feared losing power or worried that broad participation would undermine confidentiality.[55] My research investigates barriers to a more collaborative approach in the Murray–Darling Basin, examining the role of experts, power dynamics and fears around including farmers in policy processes.

In model deliberative processes, participants are open to changing their opinions through persuasion. Deliberative processes are characterised by respect, sharing of information and allowing all actors to participate freely. Straight interest-based bargaining, coercion, manipulation, manufactured consent or deception is not a part of the deliberative process.[56] But all governance involves power: more powerful actors receive favourable outcomes. The question of how power dynamics can challenge or reinforce polycentric governance systems is therefore critically important. Despite its central significance, Morrison and colleagues argued that power has not been fully explored with respect to impacts on water governance.[57] The lens of power can reveal hidden causal relationships. Broader work on water governance

---

54   Ansell and Gash 2008.
55   Huitema, Mostert et al. 2009.
56   Dryzek 2000.
57   Morrison, Adger et al. 2019.

makes use of several power perspectives, including those of Foucault, Habermas, and Gramsci.[58] My own work uses Foucault's notion of the productive and disciplinary power of discourse in shaping outcomes.

Collaborative approaches are grounded in the assumption that all actors can contribute to the process and affect outcomes. Where actors have equal power, this may be possible. In the case of water governance, the actors who come together in the collaborative process are rarely equal. Governments can initiate collaboration in ways that serve their own needs. Private firms may seek collaborative processes to maintain their social licence to operate. Environmental organisations seek sustainability objectives, and citizens may seek to address local issues. The motivations of these actors differ substantially, and some have more power to influence the collaborative process than others. The motivations of the state may also be fragmented, with individual actors framing problems in different ways. Nonetheless, as in administrative rationalism, states remain the dominant decision-making authorities.[59] According to Brisbois and de Loë, water problems are progressively being addressed using collaborative approaches to governance.[60] Despite the trends towards collaborative governance, governments continue to play critical roles in initiating collaboration, providing financial support, and approving and implementing policies. From two case studies in Ontario and Alberta, Brisbois and de Loë found that provincial governments exerted power from agenda setting to implementation in response to socio-economic factors, reinforcing existing power structures. The position and orientation of the provinces challenged the potential for collaboration in working towards positive social and environmental outcomes. These cases point to how power dynamics can significantly affect efforts towards collaborative governance.

Other research supports the conclusion that the capacity for collaborative action is closely tied to internal power dynamics. Da Silveira and Richards looked at two cases of river basin management, one in the European Union (the Rhine) and one in China (the

---

58 See for example Behagel and Arts 2014; Zeitoun and Allan 2008.
59 Brisbois and de Loë 2016.
60 Brisbois and de Loë 2016.

Zhujiang), and found that polycentric governance was more conducive to adaptation strategies only where certain conditions were met.[61] Their work posits that the influence of a polycentric governance system on adaptive capacity depends on the internal power dynamics among the components of a system and their competitive versus collaborative patterns of interaction. In the Rhine basin, the European economic integration process allowed for a greater capacity to monitor environmental water and generate a common understanding of the problems in that basin. Communication and negotiation platforms made available by the International Commission for the Protection of the Rhine and European Union institutional structures changed how key actors perceived their individual interests. These platforms not only reduced the competition for resources but were conducive to generating an institutional environment more conducive to monitoring and sanctioning. In contrast, there was a more prescriptive use of polycentric governance in China. Managers and decision-makers were found to overstate the value of having multiple centres of power. Governance systems were fragmented, and there was difficulty coordinating to achieve shared goals.[62]

In the case of the Murray–Darling Basin, power imbalances are observed in how certain kinds of information are privileged over others.[63] While scientific evidence is critical in decision-making, a propensity towards "administrative capture" is apparent. Colloff, Grafton and Williams described a process of "administrative capture" that occurred in the Murray–Darling Basin. In this process, publications and public comments were controlled by decision-makers' contracts, intellectual property rights and control over what information is allowed to be publicly available. In some cases, threats to withdraw funding were also made. Colloff, Grafton and Williams observed that scientists trusted by decision-makers tended to share similar worldviews, and their influence on policy was built on reinforcing each other's views:

---

61   Da Silveira and Richards 2013.
62   Da Silveira and Richards 2013.
63   Colloff, Grafton and Williams 2021.

> Administrative capture of science is, by its nature, a subtle and insidious process where the scientist may be unaware of how agenda setting and the development and questioning of hypotheses is gradually reframed to conform to the policy priorities and agendas of government agencies.[64]

According to Colloff, Grafton and Williams, a shift in the stance of scientists is important because, in implementing the Murray–Darling Basin Plan, where uncertainty is high, and values are contested, science can do little to help reach a consensus or achieve a common course of action. Under such difficult conditions, policy practitioners will use science to position the debate in ways that suggest a dominant set of widely shared values.[65] Colloff, Grafton and Williams argued that the required response should be to reconcile competing values through negotiation and compromise, not science that supports policies favoured by those in authority. Authorities may facilitate research partnerships that are only open to scientists who accept the existing policy frameworks. Under these circumstances, criticism and dissent are crushed, directly suppressing contrary science outputs and self-censorship by scientists who generate such findings. Science initiatives have been eliminated if considered politically inconvenient, sometimes with negative personal and professional impacts on researchers.[66] In some cases, scientists who speak out have had their work publicly criticised.[67]

It is evident from these examples that, despite efforts towards collaborative governance, power dynamics matter. Discourse analysis offers a way to unpack how prevailing worldviews impact power dynamics in water management. From a water governance perspective, this is important because polycentric governance structures are affected by the power dynamics at play among competing actors. Discursive power focuses on the ability of actors to shape social norms, values and identities. Dominant cultural framings may prevent actors from

---

64   Colloff, Grafton and Williams 2021, 135.
65   Colloff, Grafton and Williams 2021.
66   Marlow 2020.
67   Colloff, Grafton and Williams 2021.

recognising that a given outcome will harm their wellbeing or interest or that they can make a difference to outcomes.[68] The dominant discourses in my work offer a way to understand the relationship between how power manifests in the context of water governance. Wyborn and colleagues similarly drew on Dryzek's discourses of environmental problem-solving, specifically administrative rationalism, economic rationalism and democratic pragmatism, to characterise different logics of water governance in the Murray–Darling Basin, characterising these discourses as normative because they point to how environmental problem-solving should be done. These discourses also invoke and reproduce patterns of power relations. These discourses provide a way of understanding how competing institutional logics drive different approaches to governance. In administrative rationalism, actors address water governance through administrative and technical logic. In polycentrism, administrative rationalism is evident where rules, legislation and arrangements that privilege technical expertise are evidenced. Problem-solving in this logic is done through trade-offs, negotiating competing interests and enforcing agreements and rules. Power is generally constrained to central governing bodies. In the polycentric structure, efforts are directed towards creating new rules, improving the implementation of existing agreements or guaranteeing compliance. Though the approach is structurally polycentric, it does not allow for the flexibility and collaborative learning to enable adaptive governance.[69] In the context of the Murray–Darling Basin, it is important to ask how efforts towards collaborative governance are affected by administrative rationalism as a dominant discourse. The productive and disciplinary effects of this discourse on policy development, particularly collaborative actions, are explored in this work.

Democratic pragmatism, it seems, represents a positive approach to adaptive management within the context of polycentric governance structures. For Dryzek, under the logic of democratic pragmatism, "the relevant knowledge cannot be centralised in the hands of any individual or any administrative state structure ... problem-solving

---

68   Gaventa 1982.
69   Wyborn, van Kerkhoff et al. 2023.

should be a flexible process involving many voices, and cooperation across a plurality of perspectives".[70] In democratic pragmatism, new ideas are more likely to emerge, and collective learning and action are more likely to take place. Democratic pragmatism, while situated in the liberal capitalist framework, still provides a logic corresponding to many of the features of collaborative governance. Democratic pragmatism appears to represent a more straightforward road towards adaptive management within a polycentric governance system.[71] In the context of the Murray–Darling Basin, this raises the question of how democratic pragmatism may hinder or progress the objectives of collaborative governance. A prominent perspective reflected in the literature is that farmers, having caused much of the environmental damage in the basin, have little to offer in terms of input into environmental policy. In these works, the value of incorporating local knowledge in the process is questioned, with some arguing that consultations were expensive and ineffective. For instance, Crase, O'Keefe and Dollery argued that Australian taxpayers would be worse off as a result of the community consultations undertaken by the Murray–Darling Basin Authority. Garrick, Whitten and Coggan believed that the costs of consultations were too high and that if governments had limited public consultation and pursued their agenda of market acquisition unimpeded, then there would have been much lower transaction costs associated with the reform process. Further, Ross, Buchy and Proctor argued that consultations were problematic and risks were high, such as "consultation burnout"; the capacity to "raise unreasonable expectations"; the possibility that the most powerful stakeholders shape the issues, thereby limiting input from less powerful stakeholders.[72] It is important to ask, therefore, if democratic pragmatism opens up avenues for increased participation, or do the limitations associated with the discourse impede more collaborative governance processes?

---

70 Dryzek 2013, 100.
71 Wyborn, van Kerkhoff et al. 2023.
72 Crase, O'Keefe and Dollery 2014; Garrick, Whitten and Coggan 2013; Ross, Buchy and Proctor 2002, 216.

Economic rationalism sits somewhere between administrative rationalism and democratic pragmatism. Markets are constructed through administrative rules, but actors within a polycentric system have the capacity to self-organise within the market parameters and the rules defined by governments. In theory, conflicts and trade-offs are resolved through market mechanisms in which supply and demand typically allocate limited resources to the most efficient, highest-value use.[73] In practice, however, allocation decisions tend to favour those with greater economic power or political influence.[74] The economic rationalism logic does not intend to address political disparities or give voice to economically disadvantaged people. The influence of economic rationalism in shaping how power is distributed through a governance system is, therefore, an essential consideration for adaptive water governance.[75] In Chapter 3, my work examines the relationship between economic rationalism and collaborative governance. For example, do market mechanisms give agency to farmers in the Murray–Darling Basin and encourage more active participation? Conversely, does a more market-based system contribute to power imbalances that interfere with collaborative processes?

Polycentric arrangements are often considered normatively desirable, given their capacity to enable transformative change through their perceived adaptive capacity.[76] But the benefits of polycentric governance systems are questionable, given the overarching logic within which these systems operate. Examining power dynamics within polycentric governance systems can offer us a better understanding of how water governance systems can thrive or fail. This examination can also help us understand the conditions and tools required for introducing non-hierarchal and more bottom-up governance structures. An analysis of Dryzek's dominant environmental problem-solving discourses shows that some forms of polycentricity are likely more adaptive and transformative than others.[77] Chapters 3 to 5 explore

---

73   Wyborn, van Kerkhoff et al. 2023.
74   Alexandra and Rickards 2021.
75   Alexandra 2019.
76   Pahl-Wostl 2020.
77   Wyborn, van Kerkhoff et al. 2023.

the important intersection between power, discourse and governance decisions. Discourses have important productive and disciplinary effects that affect the adaptive capacity of environmental systems. Despite efforts towards polycentric water governance structures, these power dynamics intervene and shape the courses of action that are taken.

Much of the literature on polycentric governance looks at how governments attempt to engage stakeholders in deliberative processes.[78] Some attention is also paid to the ways governments privilege certain types of information.[79] Further, some scholars have examined discursive factors that may limit the capacity for polycentric governance. Wyborn and colleagues examined 34 reviews of water governance in the Murray–Darling Basin since 2004, seeking to identify systems that are resistant to change and maintain conventional discursive paradigms, and systems which enable adaptive transformation.[80] But there have not been attempts to characterise how a specific group of stakeholders, like farmers, construct their understanding of the issues. This detailed discursive analysis situates the voices of farmers as central to understanding the conditions under which they may be engaged in the collaborative process. The analysis of farmer discourse in this book offers insight into how those affected by policies understand the impacts of these policies on their lives. Resistance to the dominant discursive framings of an issue is better understood by uncovering the assumptions and underlying motivations of individual actors affected by policies. For governments to reconcile competing interests and values, they must understand the underlying motivations of actors. For example, it is unwise to assume that stakeholders are motivated solely by economic interests, or that they are unconcerned about environmental interests. They may have interests that extend beyond economic considerations, and their understandings of the environment – and their relationship to the environment – may influence their understanding of what actions should be taken to deal with environmental issues. Understanding how

---

78 Brisbois and de Loë 2016; Da Silveira and Richards 2013.
79 Colloff, Grafton and Williams 2021.
80 Wyborn, van Kerkhoff et al. 2023.

individual actors rationalise and represent their goals can provide an alternative perspective and an entry point for engaging more constructively in deliberative processes. This bottom-up approach to understanding processes of deliberative governance, rooted in a discursive analysis, offers a novel approach to research on environmental governance. Power dynamics are recognised as a major contributing factor in the level of effectiveness in developing collaborative approaches.[81] But to address power imbalances, we must understand how individual actors are empowered or disempowered by processes of government engagement.

A theoretical introduction to five environmental discourses

John Dryzek identified several "discourses of environmental problem solving", including administrative rationalism, economic rationalism and democratic pragmatism.[82] Dryzek also discussed several other "green" discourses in his work. The fourth discourse I have included here is green environmentalism, based on these other green discourses, as well as a literature review of environmental politics writers. This discourse was developed based on the observation that green discourses shared a common "ecocentric" orientation.[83] Integrating Mehta's approach to ideas, this section explains how these four discourses entail certain public philosophies, problem definitions and policy solutions, focusing on the relationship between problem definition and policy solutions. Further, building on the work of Dryzek (and his understanding of ecological democracy), Elinor Ostrom, Murray Bookchin and others, and findings from discussions with farmers, a fifth discourse called community-centrism is developed. It is distinguished from other discourses by the way it conceptualises human social relationships and ecology. This fifth category proposes a new set of policy solutions for the challenges facing

---

81   Behagel and Arts 2014; Zeitoun and Allan 2008.
82   Dryzek 2013, 73.
83   Naess 1989; Shiva 1989; and others.

the Murray–Darling Basin based on entirely different problem definitions.

Problems with the Murray–Darling Basin are defined primarily based on assumptions about relationships to nature. For example, under administrative rationalism, if we assume that nature is subordinate to human problem-solving, we might say that the problems should be solved through technological solutions. This analysis focuses on the problem definitions of each discourse and what types of policy solutions stem from these definitions. It is important to recognise that there is overlap between the elements of these discourses. But this approach provides a basic framework for understanding how problem definitions and policy solutions interact, including the disciplinary and productive effects of problem definitions. This analysis shows that "resistance" most often occurs at the levels of problem definition and policy solutions.[84] While there are numerous discourses of the environment, the five discourses described in what follows are used based on their relevance to the Murray–Darling Basin.

This first discourse of environmental problem-solving, administrative rationalism, focuses on the role of government-delegated experts in responding to environmental problems.[85] Regarding public philosophy, hierarchal social relations lie at the heart of administrative rationalism. Ontologically, administrative rationalism assumes a view of the world in which nature is subordinate to human problem-solving. Nature is viewed as manageable and amenable to human interventions. Epistemologically, administrative rationalism assumes that we can shape environmental outcomes to our benefit through strong management grounded in scientific expertise. Nature is part of a hierarchal structure with human beings at the top. As John Dryzek explained, in addition to the conception of nature as hierarchal, there are two other kinds of hierarchy at work in this discourse: one that subordinates people to the state and one in which the experts and managers have dominant positions within the state's hierarchy.[86] Torgerson and Paehlke wrote: "The image of the administrative mind is one of an impartial reason

---

84 Stone 1989.
85 Dryzek 2005.
86 Dryzek 2013, 89.

exercising unquestionable authority for universal wellbeing; it is an image that projects an aura of certain knowledge and benign power".[87]

Problems under administrative rationalism are solved through rapid modernisation or progressive reforms under the guidance of those the state deems to be expert authorities. The discourse is also characterised by steadfast confidence in science and technology to reorder the social and natural world. Further, it focuses on the overarching goals and pre-eminence of the state. A productive effect of this discourse on solutions is the prioritisation of large-scale infrastructure projects developed at a rapid pace.

Administrative rationalism assumes that power is and should be hierarchically organised. The discourse has become institutionalised in a way that there are hierarchal structures, rules and resources that support it. These institutionalised structures have the disciplinary effect of narrowing the voices that are heard and limiting alternative development visions. Powerful special interests like corporate lobby groups can more effectively interfere with decision-making, because they have the resources and connections to reach decision-makers. Information is considered valuable when it is legitimised through these institutionalised structures. While there is strong potential for problem-solving under administrative rationalism, policy definitions and solutions are limited by hierarchal power structures. Power dynamics are hidden from view because there is an assumption that decisions are informed only by expert analysis and data, not interests. These dynamics can result in their own implementation deficit, such as when hidden power dynamics effectively hinder or subvert decisions.[88]

Administrative rationalism can be identified by its focus on professional resource management bureaucracies, central agencies, regulatory policy instruments, expert advisory commissions and rationalist policy analysis techniques. Further, there are transcripts that help identify administrative rationalism including "science-based decision-making", "expert management", "risk-based analysis", "technological innovation" and "public interest".[89] Chapter 3 explores

---

87   Torgerson and Paehlke 2005, 279.
88   Dryzek 2013, 93.
89   Dryzek 2013, 75–98.

these transcripts in detail, with particular attention to how these transcripts are reinforced, challenged and subverted.

Over the 20th century, administrative rationalism allowed for the development of some of the most advanced hydrological systems in the world. Still, centralised planning also led to the centralisation of power. Further, power dynamics could be hidden from view by the appearance of rational expert analysis, which can effectively silence dissenting opinions.

The second discourse is economic rationalism. Economic rationalism as a discourse prioritises individualism in addressing environmental issues. This discourse reflects an alternative public philosophy that prioritises private over public responsibility. Economic rationalism relies on and reproduces the notion that decision-making should happen at the individual level.[90] Ontologically, while administrative rationalism stresses the role of the government and its designated experts, economic rationalism emphasises the importance of the individual. The tenets of economic rationalism became popularised during the 1970s oil crisis and the global recession that followed. Before the crisis, administrative rationalism was the predominant discourse; centralised bureaucratic planning, massive modernisation projects under state leadership and reliance on experts were the order of the day. The oil crisis, followed by skyrocketing inflation, led to a push for smaller governments, greater reliance on "free market" economies, a stronger focus on property rights and financialisation. Each of these developments accelerated through the 1980s.

The term "economic rationalism" pre-dated the oil crisis and became widely used in Australia in the early 1970s.[91] By that time, there were already calls in Australia for free market reforms in agricultural and water markets. The term was used by those who opposed tariffs and agricultural price support schemes and favoured market reforms like water trading. Epistemologically, the economic rationalists presented their policies as resting on unbiased facts that would serve the best interests of society. Opponents of the free market approach were

---

90   Dryzek 2005.
91   Stokes 2014, 195.

characterised as "irrational", basing their decisions on belief or adherence to the status quo.[92] Economic rationalism in Australia was influenced by events in the United Kingdom and the United States, where British Prime Minister Margaret Thatcher (1979–90) and US President Ronald Reagan (1981–88) led the drive for free market capitalism and conservative social policy.[93]

Economic rationalism is based on the premise that individualism promotes competition and economic growth. Economic incentives can influence individual behaviours in targeted ways. Price-based incentives are generally seen as sufficient to change individuals' behaviour. In the case of environmental management, tensions exist between economic rationalism and administrative rationalism as the dominant political discourse. Economic rationalism posits that free markets should be the dominant method of decentralising environmental planning and that markets are the most reliable mechanism for managing common pool resources. The privatisation and marketisation of public goods like air and water are part and parcel of this approach. Privatisation and commodification are the "rational" approach. Sometimes the role of the administrative state is undermined by the economic rationalist model as states are no longer in complete control of common pool resources. This view relates to the work of economist Friedrich von Hayek, best known for his defence of free market liberalism,[94] which highlights individual economic responsibility as the primary way of defining and addressing problems.

Within the environmental literature, Garrett Hardin's work on the tragedy of the commons warns of the dangers of the open use of common pool resources.[95] Hardin's work is frequently cited by those who wish to show that privatisation is the only reliable form of effective environmental management.[96] Economic rationalists argue that free markets and the protection of individual property rights are the only

---

92 Quiggin 1997.
93 Stokes 2014.
94 Hayek 1944.
95 Hardin 1968.
96 See for example Adler 2000; Anderson and Leal 2001; Anderson and Libecap 2014.

way to prevent the tragedies associated with state-centric environmental planning. Proponents say that economic rationalism avoids the tragedy of the commons situation because private property is something that property owners are more likely to care for than public or common property.[97] The disciplinary effect of this discourse is to limit the range of policy solutions to market-oriented ones, which often means privatising common pool resources. But managing collective resources under private rights regimes is challenging because, coincidently, the level of interventions required often elevates the need for strong government.[98] Another limitation of economic rationalism is that it can undervalue the multifunctionality of ecosystems by focusing only on economic outcomes in the short term, with less attention to other values, particularly social and environmental values.[99]

The productive effects of economic rationalism, contrary to arguments made by many liberal economists, are not always in the collective interest. While liberalisation may sometimes lead to better stewardship of the land, many liberal economists ignore that one effect of privatisation is that only the privileged few will be able to enjoy using that land. Further, the focus on individual land use fails to acknowledge that every land entitlement is part of a larger system, the functionality of which is contingent on the individual parts working together. For example, one person could optimise their water use by building a reservoir on their property. Every property downstream will in some way be affected, either positively or negatively. In many cases, a policy focus on economic rationalism may allow for greater efficiencies but, when governments operate within this discourse, unintended productive effects are not fully considered.

Economic rationalism can be identified by its use of several metaphors, transcripts and rhetorical devices, including "freedom", "competition", "risk management", "market-based solutions" and "efficiency".[100] These transcripts will be explored in detail in Chapter 3, as well as the assumptions reinforced through economic rationalism and

---

97 Hardin 1968.
98 Robertson 2007.
99 Hollander 2007.
100 Dryzek 2013.

its productive effects. Such an analysis allows us to see how this discourse hinders the capacity of water management policy to adapt to some of the challenges it faces.

The third discourse we are concerned with is that of democratic pragmatism. Democratic pragmatism as a discourse emphasises democratic processes and interactive problem-solving among participants. In the Western world, policy development is typically centralised, so decision-makers are often far removed from how human–environment interactions play out at the local level. Decentralised and participatory decision-making represents an alternative approach. The epistemological assumption underlying this discourse is that knowledge comes from different citizen participants, entailing a more bottom-up approach to policy formation. There are numerous reasons why a more participatory approach is desirable. From an institutional perspective, for example, participation can be used to attain a goal defined by someone external to the community involved. For social movements, participation is a goal as it is an empowering process that can also help define a community's goals.[101]

John Dewey, an American who advocated democratic pragmatism in the 1920s and 1930s, believed that participation was the most fundamental aspect of democracy, not representation. This idea underpins democratic pragmatism as a discourse. For Dewey, democracy could only be accomplished through active communication between citizens, government officials and experts.[102] Similarly, the European critical theorist Jürgen Habermas argued that government policies and legal instruments must allow for the personal autonomy of those subject to them.[103] Legally enshrined rights can only offer citizens true freedom when those citizens can exercise public autonomy. Ontologically, democratic pragmatism is fundamentally individualist, like economic rationalism. But, in democratic pragmatism, the focus is on the individual as a political actor who is part of a larger collective. Democratic pragmatism asserts that the rights that define individual freedom must also include the rights of political participation. As

---

101 Tufte and Mefalopulos 2009.
102 Dewey 1916.
103 Habermas 1981.

Habermas understood the relation between private and public autonomy, each conceptually presupposes the other in the sense that each can be fully realised only if the other is fully realised. The exercise of public autonomy in its full manifestation presupposes participants who understand themselves as individually free (privately autonomous), which presumes that they can shape their freedoms through the exercise of public autonomy. Habermas summarised this requirement in his democratic principle of legitimacy: "only those statutes may claim legitimacy that can meet with the assent of all citizens in a discursive process of legislation that in turn has been legally constituted".[104] Both Dewey and Habermas argued that democracy wholly depended on the freedom of the individual to assert their influence in the public realm. Democracy, in any true sense, depends on individual participation – which allows people to experience personal autonomy first and foremost.

In the tradition of Habermas and Dewey, environmental theorists have argued that complex environmental issues can only be solved through deliberation and public participation.[105] Hajer, for instance, presented a new perspective – "ecological modernisation" – which stresses the opportunities of environmental policy for modernising the economy and argued that this more collaborative approach of the 1990s and onwards replaced the antagonistic environmental debates of the 1970s.[106] Democratic pragmatists argued that environmental problems could only be resolved by integrating the various perspectives of a wide range of individuals and groups. The discourse can thus be seen as a way of decentralising and expanding the problem-solving community. Problems are defined in a way that assumes that they can be resolved within the system, so long as there is enough active participation from the communities involved. Therefore, the epistemology of the discourse assumes that knowledge produced through consultative, participatory processes informed by myriad individuals of diverse positionalities (notably including scientists and other types of experts) is better than that which a few people produce or is only reliant on the scientific

---

104 Habermas 1991, 110.
105 Dryzek 2005.
106 Hajer 1997.

method. Unlike administrative rationalism, the discourse of democratic pragmatism treats the government as a system of interacting processes that citizens are entrenched within. Like administrative and economic rationalism, democratic pragmatism places nature as subordinate to human problem-solving efforts. Whether nature is a commodity or a self-regulating system is of little consequence for democratic pragmatists who believe deliberation between individuals will solve conflicts in the system.[107]

There are at least two significant problems associated with democratic pragmatism, both in theory and practice. First, democratic pragmatism treats participation as synonymous with having a role in decision-making. This conflation is problematic because it ignores other relations of power that may affect how decisions are made. Participatory models that try to incorporate local or experiential knowledge into the existing institutional structures often reproduce government officials' views and enhance state power. This occurs because knowledge as a process of learning develops within cultural, political and economic contexts. For local knowledge to be incorporated into decision-making, it must first meet the criteria set out by the state as to what constitutes valid knowledge. The devaluing of certain types of knowledge has disciplinary effects on the policy reform process. While proponents of democratic pragmatism seek to incorporate alternative views into decision-making, the experiences of local people are often watered down or dissected into parts to fit the established narratives of the bureaucracy best. Democratic pragmatism can thus be limited by the other dominant discourses, particularly administrative rationalism.

Second, in practice, democratic pragmatism is often not truly participatory because it ignores power dynamics among participants, overlooks how legitimate knowledge is constructed and discounts the ways that human beings are often at the mercy of processes in nature. In his work on deliberative governance, Dryzek explained that a system can be said to have deliberative capacity if it has structures to accommodate deliberation that are *authentic, inclusive* and *consequential.* To be authentic, "deliberation ought to be able to induce

---

107 Dryzek 2013, 111, 114, 115.

reflection upon preferences in noncoercive fashion". To be inclusive, "deliberation requires the opportunity and ability of all affected actors to participate". Finally, to be consequential, "deliberation must somehow make a difference when it comes to determining or influencing collective outcomes".[108] Democratic pragmatism, in practice, often fails on one or more of these fronts.

The discourse of democratic pragmatism is often associated with individualism, focusing on the role of individuals participating and being consulted. Democratic pragmatism is easily identified by specific transcripts, particularly among government agents. An emphasis on "deliberations", "consultations", "committees", "engagement strategies", "education" and "reason" characterise the discourse.[109] It is important to draw this discourse out as a unique category because there are distinct productive effects of democratic pragmatism.

Several alternative approaches to environmental decision-making are discussed by Dryzek, including deep ecology, ecological modernisation, eco-anarchism, green consciousness, eco-theology, bioregionalism, sustainable development and social ecology. Dryzek did not present these as a single discursive category but referred to them as variations of green consciousness. In this work, I have constructed a category labelled green environmentalism, which includes elements of many of the above perspectives.[110]

In the Western world, the distancing of people from nature was seen as necessary for the proper functioning of society. This view has profoundly shaped interactions with the natural world, in that governments, corporations and institutions tend to see nature as separate from people. The effect of a discourse that prioritises culture (that which people produce) over nature has tended to diminish our perceptions of the extent to which human activity has environmental consequences and the extent to which natural environments influence human affairs. While green environmentalism is a healthy reaction to

---

108 Dryzek 2010, 10.
109 Dryzek 2013.
110 Note that in this work social ecology is included as part of a separate category, community-centrism, the fifth discourse relevant to the examination of the Murray–Darling Basin.

anthropocentrism, it is problematic because it ignores the already well-entrenched and fragile ecological relationships between people and the environment resulting from our historical legacy of interference in natural systems.

In contrast to this dualistic and hierarchical view, in many older societies, nature is viewed as a sentient force constantly acting on human beings. In Indian cosmology, all existence arises from the primordial energy, which is the substance of everything, pervading everything. The manifestation of this power, this energy, is called nature (Prakriti).[111] These holistic views of nature are internalised within social discourse in meaningful ways. In contrast, Western societies tend to resist this kind of internalisation as superstitious, even when scientific evidence supports the assertion that human-produced injury to natural environments has detrimental consequences for people. Measures to address environmental problems in Western societies tend to focus on creating additions to existing policies but do not give much consideration to the wider system in which these policies are being implemented. Green environmentalism applies romanticised assumptions about nature, wherein humans are separate from and perceived as in conflict with natural processes.

Arne Naess, a Norwegian environmental philosopher who coined the term deep ecology, saw humans in conflict with nature. He advocated for biocentrism (the belief that all forms of life are of equal value to human life) and argued that people need to identify with a larger self beyond that of the individual (which he termed "self-realisation").[112] Naess was reacting to the ecology movement at the time, which sought only to moderately reform the environmental practices of the industrial age to benefit human beings.[113]

The ontological stance of green environmental discourse is thus premised on a representation of human beings as a threat to natural environments by virtue of their domination, which presents difficulties and limitations in terms of environmental policy. It privileges green activism and a wilderness perspective wherein we see nature needing

---

111 Shiva 1989.
112 Naess 1989.
113 Dryzek 2013, 187.

protection from human beings. This narrative has effectively drawn much needed attention to the movement and the detrimental impacts of unfettered development. In the 1970s, for example, environmentalists in Australia focused much of their energy on the Great Barrier Reef, hoping to protect it from mining and oil drilling. Community-driven support for the reef's protection garnered national attention, and the Commonwealth government established a Marine National Park and Management Authority in 1981. Similarly, a campaign against a dam that would flood the valley of Tasmania's Franklin River mobilised tens of thousands of Australians in the late 1970s and early 1980s and helped solidify the base of Australia's environmental movement. At the same time, the knowledge of local people trying to manage the land can be undervalued. This is because, epistemologically, the discourse – as developed in the modern West – generally puts faith in science over traditional or local forms of knowledge.

Green environmental discourse can be identified in a variety of common transcripts, including terms like "wilderness", "natural", "unnatural", "protection" and "environmental values". These transcripts will be explored in detail in Chapter 4. That chapter focuses on the dichotomy between natural and unnatural environments. I look at how "wilderness" and "nature" exclude environments with human production activities, how climate and weather are understood and represented in the green discourse, and why these transcripts are often problematic.

A fifth environmental problem-solving discourse, community-centrism, is introduced in Chapter 5. Community-centrism is inspired by the writings of critical environmental scholars like Bookchin, Ostrom and Li, and scholarship on community-based adaptation to climate change. This scholarship can help us understand the contours of an alternative environmental problem-solving discourse that puts the needs and capacity of local communities front and centre in defining solutions to their challenges.

Community-centrism is influenced by social ecology, which was introduced by the anarchist author and environmental philosopher, Murray Bookchin. While it does not ignore the central importance of economics or environmental management, social ecology points to the overwhelming significance of human social relationships in determining economic and ecological outcomes. Bookchin argued for an environmental politics that places human social interactions as central. He believed that the roots of ecological problems are closely tied to human social problems and can be solved by reorganising society along more ethical lines.[114] His approach acknowledges the co-dependent relationships between human communities and natural systems. Bookchin argued that environmental problems are a product of the degradation of human beings by hunger, material insecurity, class rule, hierarchal domination, patriarchy, ethnic discrimination and competition.[115] Bookchin traced the roots of environmental problems to social problems and advanced his brand of libertarian socialism. He proposed that empowering people at the local level and giving them the democratic tools to engage fully within their communities invariably results in positive economic and environmental outcomes. Since the destruction of ecological systems can be traced to social systems of domination, Bookchin asked us to replace these systems with participatory democracy at the level of communities. Essentially, by changing how we relate to one another, we change how we relate to our environment.[116] Ontologically, community-centrism places the role of the community (or society) as central.

Like Bookchin, Elinor Ostrom focused on human societies' role in positively affecting their environments. In environmental resource management, this is done through gaining knowledge of locally specific contexts and acting according to principles that meet the needs of local communities. While Bookchin advocated for a more radical approach to politics, Ostrom worked within the confines of established frameworks for ecological management to define eight clear principles for managing common pool resources. Among the eight principles, she argued that the commons need to have clearly defined boundaries (particularly around who benefits), that rules should fit local circumstances, that participation is critical, and that the commons must have the right to organise.[117] Likewise, community-centrism discourse

---

114 Bookchin 1994, 1982.
115 Bookchin 1994.
116 Bookchin 1982.

focuses on participation, local circumstances and the capacity to manage the commons collectively. The problem definitions and policy solutions that arise from this discourse reflect these principles. The principles of common property resource management are widely accepted and used today. Ostrom has won global praise for her work, including the Nobel Prize in Economics, the first awarded to a woman.

Like Ostrom, Tania Murray Li argued for an approach favouring local decision-making where possible. She also made the case that the relationship between local and expert knowledge can be reciprocal in such contexts.[118] While Li's work was critical of the capitalist superstructure that tends to define the relationships between government experts and agrarian landholders, she saw many opportunities to transform that relationship within the existing system.

While there are many differences between these thinkers in terms of the overall focus of their work, for Bookchin, Ostrom and Li, some of the most effective solutions to environmental management arise out of policies that support collective accountability on the part of all actors and allow land managers like farmers to play a central role in the planning process. Epistemologically, community-centrism focuses on the role of local knowledge and knowledge exchange. Farmers have a rich tradition of sharing ecological knowledge to produce alternative policy prescriptions and outcomes.[119] They also employ a wide range of tools to respond to the demands of local ecological systems, including multiple species management, resource rotation, revegetation and erosion control to name a few. Farmers form social mechanisms for transferring this knowledge within their communities, which helps them to maintain traditional practices and learn from one another. Farmers are constantly responding to the real-time demands of their environment and are thus guided by experience and social networks of knowledge transfer more than other types of knowledge. In other words, policy solutions often reflect the significant role of social networks at the heart of effective ecological·management. Problems are defined in a way that highlights the role of community support and

---

117 Ostrom 2012.
118 Li 2007.
119 Berkes, Colding and Folke 2000.

people's social networks as critical components of managing ecological systems.[120] Transcripts include such terms as "community-based solutions", "bottom-up", "local knowledge" and "localism". Community-centrism can be characterised as a discourse of resistance.[121] How this discourse looks in practice, in terms of how community members (especially farmers) characterise their approach to addressing the problems of the Murray–Darling Basin, is explored in Chapter 5.

Table 2.2 builds on this theoretical framework and the preliminary introduction to the five discourses. It summarises key problem definitions that relate to each discourse and how these definitions interact with policy instruments. It is important to acknowledge that problem definitions are not necessarily confined to just one discourse. Still, the table provides a conceptual framework for understanding the main differences in these approaches and their respective implications.

This book explores how discourses shape policy solutions for dealing with the significant challenge of effective water management in the Murray–Darling Basin. Only through an understanding of the ways that governments and farmers define the problems can we understand how policy solutions develop. Collectively, these five discourses offer a way of understanding the political dynamics in the Murray–Darling Basin. The analytic framework connects the way problem definitions are connected in discursive terms to the policy solutions employed in the Murray–Darling Basin. Through this analysis, we can determine how discourses shape possibilities and limit the capacity to develop policy responses. Further, reframing problem definitions may offer new ways of understanding and addressing water management problems.

---

120 Berkes 2009.
121 Foucault, 2002.

Table 2.2 Discourses, problem definitions, policy solutions, transcripts, metaphors and rhetorical devices

	Discourses	
Problem definitions	Types of policy instruments	Transcripts, metaphors and rhetorical devices
	Administrative rationalism	
Science and technology should be primary tools for analysis	Professional resource management bureaucracies	Expert management
Focus on risk-based analysis and adaptation measures	Central agencies	Science-based
Reliance on bureaucrats, scientists and engineers is central to decision-making	Regulatory (command-and-control) policy instruments	Public interest
Cost-effective measures a key priority	Expert advisory commissions	Risk-based analysis
		Technological innovation
	Economic rationalism	
Economic growth and competition are inherent goods to be protected from unnecessary interventions	Market-based instruments	Freedom
Individual property rights should be protected	Privatisation	Competition
Free market mechanisms can be relied on	No government intervention	Risk management
Government interventions are undesirable	Marketisation	Market-based solutions
	Asset management	Efficiency
	Commodification of water	

Discourses		
Problem definitions	Types of policy instruments	Transcripts, metaphors and rhetorical devices
Democratic pragmatism		
Democracy through participation Communicative democracy and rationality central (Habermas) Environmental problems are solved through participation (Dryzek, Hajer)	Consensus-building conferences Opinion polls Town hall meetings	Citizen science Deliberation Consultation Engagement strategies Education Reason
Green environmentalism		
Nature needs to be protected against people/farmers Privilege wilderness over other environmental spaces Mystification of nature and reverence for ecological systems outside human intervention Humans should be considered separate from nature Social problems are subordinate to ecological problems	Regulations to protect nature from people Protected area management Government interventions Expert and government management	Protection Wilderness Natural/unnatural Environmental values

Discourses

Problem definitions	Types of policy instruments	Transcripts, metaphors and rhetorical devices
	Community-centrism	
Social challenges are the cause of environmental problems	Locally based planning committees	Community-based solution
Strong social relationships are essential for environmental management	Farmer-driven environmental initiatives	Bottom-up
Local knowledge helps solve problems	Community-oriented social welfare programs	Community
The relationship between local and expert knowledge should be reciprocal	Extension services that focus on local growing conditions	Local knowledge
Social, historical and geographical contexts are key considerations	Community-based hubs of knowledge exchange	Localism
Resilience through self-management		

75

A note on methodology

As incidences of drought and flooding increase throughout the world, it is necessary to understand how policy responses may develop in different contexts and why they may fail or succeed. An in-depth analysis of the Murray–Darling Basin provided an ideal opportunity to explore the discourses and policy processes that can contribute to environmental crisis, information that can be used to better understand similar situations. A single case-study approach produces a context-dependent knowledge that helps develop a deeper understanding of an issue.[122] The case of the Murray–Darling Basin provides critical insight regarding the role of farmer knowledge in policy development. Farmers are constantly responding to the real-time demands of their environment and are thus guided by experience and social networks of knowledge transfer. Community networks of farmers are a key component of how ecological systems are managed.[123] Ecological knowledge is formed and reproduced through the daily interactions that farmers have with their environments, and through the social networks that they have formed with each other over the course of multiple generations. As such, this knowledge is critical in understanding how water is managed and what farmers can contribute to address problems.

The research process involved four parts: document analysis; semi-structured interviews with farmers and government officials; other data collection; and data analysis.

One of the first tasks was to define and describe the dominant policy discourse informing the management of water resources and to consider what this discourse means for Murray–Darling Basin farm communities. This task involved careful analysis of the specific policy instruments used to manage water in the basin. The three most prevalent policy instruments identified were environmental buybacks, water trading (market-based instruments) and drought assistance (including adaptation and mitigation measures). Each of these

---

122 Flyvbjerg 2006.
123 Berkes 2009.

instruments has significant implications for farmers, but I hoped to understand how farmers informed the development and implementation of these instruments, if at all. I also wanted to understand how farmer involvement was framed: were farmers included in the development of these instruments? How were farmers represented: were they regarded as part of the solution or part of the problem? Further, given the market-oriented approach of policies in the Murray–Darling Basin, how were farmers valued within the system?

I asked farmers and government officials questions meant to elicit a better understanding of how the three main policy instruments, described above, functioned in the basin. I also asked farmers questions that would help me understand their views on the Commonwealth and state governments' management of the Murray–Darling Basin. I asked questions concerning the ecological knowledge of farmers and how this knowledge has affected water management policies. I asked them what sources of information were important to them and where they gathered knowledge. Their answers helped me understand whether the information they relied on was locally based ecological knowledge, knowledge collected from the scientific community, or both. I also asked how they saw their role and ability to positively respond to the challenges presented to them, and what social norms or expectations influenced their ability to act. I undertook 34 semi-structured in-person interviews: 25 with farmers, eight with government officials and one with a representative of an irrigators' council.

In my initial research design, I used these selection criteria for respondents: the farmers I interviewed depended primarily on farming for their livelihoods, were using irrigation to farm (or had used irrigation in the past and transitioned away from irrigation farming), and lived and worked around the Murrumbidgee and Murray catchment area. Most of my respondents were in the region that includes Narrandera, Leeton, Finley and Griffith. I also did a few interviews near Albury, on the New South Wales–Victorian border, and one farmer interview in Victoria. There are significant differences in the kinds of farms that operate in the Murray–Darling Basin. The farms ranged in size from 800 to 1,600 hectares. I did not limit the types of farmers in the project because I was interested in gathering a range of perspectives from different farming operations, but most farmers in my sample grew rice. A grape grower and a plum producer were the exceptions. Most farmers had a diverse range of crops in rotation to better manage soils, and several integrated sheep farming within their operations. The farm businesses in my sample were family-run operations. They did not include farm operators employed by corporations (other than companies owned by their own families to organise an intergenerational business). The majority of rice farms in the Murray–Darling Basin are family-run operations with a long history of farming in the region.

Farmer interviewees were identified using snowball sampling, which involves asking interviewees for referrals. Snowball sampling led to many repeated referrals, and it became clear that the farming community was relatively small and tight-knit, despite the vast distances between the farms. Because of this, I concluded that I reached an acceptable saturation level in the variety of responses I could expect from farmers. Farmer respondents were eager to speak with me; only one farmer could not accept my interview request due to a crisis on his farm. This eagerness resulted in interviews that sometimes lasted several hours and provided me with a wealth of information to work with. I was also able to attend local meetings and community gatherings to meet with research participants.

I also interviewed respondents from the Murray–Darling Basin Authority and the National Water Commission. I interviewed eight officials located in Canberra using open-ended questions, asking them to explain the rationale for programs related to the management of the Murray–Darling Basin, to discuss the effectiveness of those programs, to offer analyses regarding how those programs and policies may be affecting farmers, and find out how farmers were included in the policymaking processes. To understand the types of considerations that inform the discussions among government officials, I also inquired about the information they relied on. For example, I asked about the information used in formulating and framing problems, what resources were available to assess the effects of the measures that were implemented, what actions were taken to enhance the collective knowledge of problems identified (surveys of local knowledge and scientific assessments), and how primary data was used in the decision-making process.

Dryzek drew on an extensive body of evidence to develop his variations of environmental discourse. My analysis began with an empirical application of these discourses. I used a selection of Dryzek's discursive categories that seemed particularly relevant to revealing the story of policy change in the Murray–Darling Basin. In particular, the discourses of economic and administrative rationalism appeared to be dominant among both farmers and government officials.

I quickly recognised recurring themes and concepts – like allocations, acquisitions, impacts of policy, on-farm system types, ecological knowledge types, and science-based initiatives – in the farmer and government interviews and developed a model for categorising responses based on the five central discourses. Even though I began with these, I quickly realised certain recurring themes were also consistent with Dryzek's environmental problem-solving discourses.[124] This led me to further explore Dryzek's discourses and how they related to my data. Some transcripts in my data did not conform to the dominant discourses or to green discourse. I grouped these ideas and practices into a new category labelled "social ecology" discourse, because of their attention to the role that people play in managing and sustaining natural environments. I then investigated the literature on social ecology and political ecology, which led to the development of the fifth environmental discourse: community-centrism.

Discourse analysis is an interpretive and reflexive research technique.[125] As Phillips and Hardy noted, the strength of discourse analysis is often in its "contextual and interpretive sensitivities", which set it apart from more traditional methods. In critical discourse analysis, validity and reliability are understood differently because researchers trace multiple meanings to understand their implications, which is a largely subjective process.[126] This discursive analysis of water politics in the Murray–Darling Basin is the result of a years-long

---

124 Dryzek 2013.
125 Phillips and Hardy 2002, 5.
126 Phillips and Hardy 2002, 75.

iterative process. Although this is an unconventional approach, particularly in political science, it allowed me to incorporate categories and details of discourses that participants constructed themselves, including the knowledge and ideas they contributed, as well as how they challenged those of the government. This approach also reflects a bottom-up approach wherein we can examine how actors construct and challenge discourses from below. This discursive interpretation is explored in Chapters 3 to 5.

Conclusion

The case of water management in the Murray–Darling Basin can be seen as an example of what Flyvbjerg calls an extreme or "critical" case,[127] given that the drought conditions in the region are severe, that the government's response has been extensive, and on-farm adaptation measures have been substantial. This research highlights how dominant discourses define policy choices and how farmers seek to redefine the policy debate. The chance to talk with farmers in their own space and on their terms gave them a sense of comfort and allowed them to express themselves freely. Even though there were numerous challenges to conducting these in-depth interviews in rural Australia, I believe this kind of research is indispensable in developing a rich and detailed understanding of the challenges faced in situations like the Murray–Darling Basin.

The discursive categories developed in this research emerged iteratively; I started with a framework found in Dryzek and added to it as I analysed the data. In this sense, my analytic process was both top-down (working with pre-existing discourse categories to understand the data) and bottom-up (building new categories as needed). The latter was especially useful for identifying the new discursive category I developed through this research. Community-centrism suggests that a bottom-up approach to policy formation may greatly enhance outcomes from formation to implementation. Both green environmentalism and community-centrism critique the current

---

127 Flyvjerg 2006.

environmental movement and provide an alternative approach to environmentalism, respectively. While these categories were developed based on well-established research, they provide a concise and productive way of understanding the problems and working towards solutions.

3

Administrative rationalism, economic rationalism and democratic pragmatism

This chapter tells the story of Murray–Darling Basin management, a story in which administrative rationalism dominates. Administrative rationalism has the goal of rapid modernisation under what is considered the expert authorities of the state. The ontological and epistemological assumptions of administrative rationalism, and its problem-solving approach (including top-down and expert-led control and decision-making), continue through each shift. This continuity in approach is critical to recognise because administrative rationalism has some fundamental limitations, manifest as disciplinary and productive effects of the discourse itself. Viewing Murray–Darling Basin management through the discourse of administrative rationalism helps explain the government's largely top-down approach to environmental water management and the effects of this approach.

Over time, and corresponding with changes in management regimes associated with neoliberalism more broadly, some of the assumptions of what Dryzek termed economic rationalism came to figure in problem definition and policy instruments enacted in the Murray–Darling Basin. There was not a wholesale adoption of a new problem-solving discourse by the state and other Murray–Darling Basin actors, as Dryzek has suggested can happen with economic rationalism in other contexts. Instead, in the Murray–Darling Basin, the adoption of market-based instruments occurred within the administrative rationalist frame. The government still led economic reforms, and regulations heavily influenced water trading. Perhaps most telling is that the Commonwealth government became the largest buyer and seller of water in the Murray–Darling Basin in the wake of reforms, further centralising government control over water. As a result, some of administrative rationalism's underlying limitations or weaknesses were simply compounded and exacerbated by the addition of these tools of economic rationalism.

Murray–Darling Basin management also had some space for democratic engagement in the form of consultations, but these were often minimal and inconsequential. The problem-solving discourse of democratic pragmatism appears to have informed the inclusion of tools and practices that engaged citizens. But, the adoption of democratic engagement generally consisted of consultations performed in a largely top-down fashion. The evidence shows that consultations did not significantly affect decision-making and were a far cry from more consequential and deliberative forms of democratic engagement that farmers and other critics of Murray–Darling Basin management believed should be in place to shape decisions.

This chapter shows that while environmental concerns of various types did rise to the fore over time, they were dealt with through assumptions related to administrative rationalism. We can see a nod to economic rationalism as demonstrated by the prioritisation of individualism over community outcomes and a nod to democratic pragmatism as displayed through the limited mechanisms for democratic influence by key stakeholders. This constellation of discursive factors was critical in creating a misalignment between what the Murray–Darling Basin management plan under the Commonwealth *Water Act* of 2007 set out to do and the environmental results for the river basin and for the human and non-human communities that inhabit it. These discourses (and their disciplinary/ productive effects) and practices (e.g., consultations, town hall meetings, stakeholder input requests) met with the worst drought in Australia's modern history to inform the creation of a new management plan for the Murray–Darling Basin.

The first section of this chapter shows how administrative rationalism came to characterise water management in the basin, examining the productive and disciplinary effects of this overarching discourse. The second section explains the emergence of economic rationalism as a complementary and competing discourse to administrative rationalism, again examining the productive and disciplinary effects of the discourse (exemplified through the emphasis on market-based policy instruments). The last section looks at the role of democratic pragmatism in shaping the consultation process in the Murray–Darling Basin. This discourse acts alongside administrative and economic rationalism to encompass the dominant discursive orientations that have shaped the Murray–Darling Basin's policy development and implementation process.

Making sense of Murray–Darling Basin management through the discourse of administrative rationalism

The issues which the Murray–Darling Basin management structures had to contend with shifted over time (from salinity issues to an increasing focus on water volumes), as did the management structures themselves (from state to more central federal government control).[1] The basic approach to addressing these issues remained largely top-down. One of the defining assumptions of administrative rationalism is that problems are subordinate to human problem-solving. Nature is considered manageable and amenable to human interventions. In its epistemological assumptions, administrative rationalism relies on scientific and technical expertise. This section first details the productive effects of administrative rationalism on Murray–Darling Basin management, including the focus on the water over-allocation problem, a reduced cap on water allocation undertaken from the top down, and the invocation of market-based instruments. Second, it introduces some unintended productive effects of administrative rationalism, explicitly focusing on water scarcity and efficiency under Commonwealth government oversight. Finally, it discusses four examples of specific disciplinary effects of administrative rationalism in the context of the Murray–Darling Basin.

---

1   Connell 2007.

Over the late 20th century, many new issues like salinity and recreational water rights came to inform basin planning, culminating in the problem of "over-allocation". The clearing of native vegetation, as well as drainage for irrigation, contributed to salinity problems. Inadequate water flows related to over-allocation prevented excess salt from being flushed down the river.[2] From 1972 to 1982, states and the Commonwealth government undertook negotiations to include issues surrounding water in the Murray, and salinity became central to defining water quality problems. In 1985 the Murray–Darling Basin Ministerial Council and the Murray–Darling Basin Commission were established (later replaced by the Murray–Darling Basin Authority). The new Commission established a water management plan among four signatory states (New South Wales, Victoria, South Australia and Queensland) and the Australian Capital Territory (ACT). Members unanimously agreed upon the resolutions of the ministerial council, and the council was required to appoint a community advisory committee and other committees as it saw fit. In 1998, a salinity and drainage strategy was introduced, including a plan for cooperation by the participating governments.[3] While the ministerial council encouraged participation among the states, authority became increasingly centralised under the direction of the Commonwealth government.

The notion of water "over-allocation" emerged within this context as a primary problem definition. From 1988 to 1993, water diversions for farming increased by more than 1 per cent annually, resulting in a cumulative increase of 8 per cent. Diversions were only 63 per cent of the total allowed under the allocation system, meaning severe water shortages would result if all allocations were used. In 1993, a working group was formed to examine and report on the scope of further water diversions and their potential impact.[4] At this time, the term "over-allocation" was first used to define the cause of water scarcity in the basin. The working group's 1993 report represented a turn on the part of

---

2   Government of South Australia Department for Environment and Water 2016.
3   Chenoweth and Malano 2001, 302, 303, 305.
4   Chenoweth and Malano 2001, 307–8.

the government from being a driver of water usage and "productivity" towards the role of the government as the central figure in imposing water reform and a conservationist approach. A later senate inquiry from 2006 described over-allocation as the critical problem in the Murray–Darling Basin, citing that 66 per cent of the water that would have normally flowed to the sea was now going towards productive uses.[5] Framing the problem of water scarcity as one of over-allocation has meant that the government has become increasingly focused on reducing water use for productive purposes.

The 1990s also saw an increased focus on using scientific data to determine acceptable diversion limits and implement a "cap". As a result of the discussions and work undertaken by subcommittees in the previous year, in 1994, a report by the Murray–Darling Basin Commission entitled *Limits to Surface Water Diversion in the Murray–Darling Basin* was released. This report was based on evidence including entitlements, allocations and state water use data. The report found that the system only limited water diversions during droughts. During non-drought periods, practices tended to encourage greater diversions. The Commission found that the licensing and allocation system needed revisions in order to decrease diversions over time. At its 20th meeting in 1996, the ministerial council introduced a cap on diversions for New South Wales, Victoria and South Australia. New South Wales and Victoria agreed to cap their diversions at 1993–94 levels of development, while South Australia agreed to a cap at a slightly higher level.[6] The decision to impose the cap was justified by scientific evidence from the Commission, working groups and subcommittees.[7] At the time, there was a general agreement among the states (as well as farmers and government experts) that a cap was needed to prevent further diversions. Many of the farmers I spoke with remembered this period as one of intense negotiations culminating in a generally accepted response to drought. But the decision to impose the cap became contentious soon after the Murray–Darling Basin Authority

---

5   Parliament of Australia Senate Standing Committees on Rural and Regional Affairs and Transport 2003.
6   Chenoweth and Malano 2001, 307–8, 309.
7   Murray–Darling Basin Authority Ministerial Council 1996.

proposed further water reforms. So, while there was a collaborative effort to negotiate the cap, what followed was a series of top-down decisions by the Commonwealth government.

Despite the cap, diversions again came to the top of the political agenda as the Millennium Drought intensified. Several wetlands, for example, gained attention internationally and were listed under Ramsar as protected sites, including the Riverland region's floodplain in the lower Murray. These sites were seen as rare examples of unique ecosystems, providing habitats to protected species and having cultural and spiritual significance for Indigenous peoples.[8] As the drought intensified, the policy solution of another cap became more attractive to policymakers. But the long process of consulting and implementing the previous cap had made many farmers hostile to water reforms. They had hoped the earlier cap would represent the end of a long process of negotiations and the beginning of a more stable period. Several farmer interviewees said they were still struggling with the impact of the reforms initiated in the 1990s, so an even more comprehensive water reform program was viewed with unease. This unease was exacerbated by the role of the Murray–Darling Basin Authority, which sought to centralise the administration of reforms borne by the states. Many farmers viewed this change as a turn towards top-down policy processes.

As explained in Chapter 1, the Millennium Drought quickly revealed the shortcomings of the established model. The Commonwealth viewed the previous cap as simply insufficient for the challenges at hand. As such, market-based instruments and environmental buybacks were introduced in 2008 as the primary policy instruments for confronting the challenge of water management. One way to understand the implementation of market-based instruments is that they were a consequence of the discourse of administrative rationalism. While market-oriented – and thus arguably also tools of economic rationalism (an argument explored below) – the market-based instruments effectively reinforced the central role of government by requiring increased government intervention. The new policy

---

8   Murray–Darling Basin Authority 2021b.

solutions presented opportunities to manage water better, but many issues appear to have been overlooked.

Two impacts of implementing market-based instruments demonstrate the unintended productive effects of the reform process: market-based instruments tended to favour large irrigators; and they emphasised scarcity and efficiency. For example, powerful large-scale cotton growers influenced New South Wales agencies to retain high water diversions. For example, in 2017, media reports suggested that opportunistic cotton growers in the Barwon–Darling catchment in northern New South Wales, in the northern Murray–Darling Basin, were misappropriating vast amounts of water needed for downstream users and environmental protection.[9] New South Wales agencies failed to close loopholes or legally enforce regulations to curb water waste and were not subject to any independent enforcement mechanism.[10] Russell James of the Murray–Darling Basin Authority corroborated this interpretation. He explained how large-scale operations, buttressed by New South Wales led agencies, came in and tended to set up the infrastructure themselves. He noted: "by and large; they're big operations, a little bit of a, dare I say, a cowboy, sort of a little bit like a western what we call it … the Wild West".[11] Despite high water-use levels, large operators had managed to avoid the regulatory reforms smaller family-run operations faced. This type of regulatory capture is not uncommon in administrative rationalism. There is a long history of large-scale irrigators co-opting governmental agencies to advance their interests. While some government officials may favour smaller-scale farms as they tended to do more to support local businesses and secondary industries, their agencies still tended to comply with the demands of larger irrigators.

A second unintended productive effect of market-based instruments was an emphasis on water scarcity and a drive towards closer government oversight. This emphasis led to increased regulations, which the government did not have the capacity to monitor or enforce. The decade-long dry conditions that began in 1997 made

---

9   Downey and Clune 2020.
10   Pittock and Connell 2010, 572.
11   R. James, personal communication, 2016.

water availability extremely low. The Commonwealth and state governments thus began to shift their focus from ensuring supply to irrigators to focusing on the environment. This period represented a significant departure from a discourse that focused on salinity towards a discourse that focused on "scarcity" and curbing "over-allocation". Ensuring adequate water supplies to the environment became a priority, and water for irrigation farmers was substantially reduced. Irrigation farmers with permanent plantings took precedence, and their water rights were generally guaranteed. Farmers who owned a water allocation could get a percentage of their water entitlement based on water availability, and any additional water could be bought at market value. Farmers who did not have an allocation could not buy water during the period due to the extreme cost. The most significant action on the part of the government at the time was the National Water Initiative, which was introduced in 2004, during the Millennium Drought. It allocated some $13 billion to deal with "water scarcity" issues.[12] A major portion of those funds was allocated to programs meant to increase water-use "efficiency". The National Water Initiative was so significant because, in the past, state governments had broadly shared power and made rules based on intensive negotiations between themselves. Instead of a shared management regime, the National Water Initiative solidified the Commonwealth government's central role through its sizeable monetary investment. The National Water Initiative included many new regulations meant to enhance water savings from an ecological and economic perspective, but the centralisation made it more difficult for farmers to access the planning process. Seen through this lens, the changes found in the National Water Initiative represented a move towards a more top-down process. Within the context of administrative rationalism, this change represented a further entrenchment of high-level government planning.

High-level management was further reinforced by the appointment of a new body called the Murray–Darling Basin Authority under the federal *Water Act 2007*. The new body was to manage the basin and report to the minister for water. The Act included an obligation to

---

12   Murray–Darling Basin Authority 2018.

develop a basin-wide plan to focus on improving the river's health and establishing a maximum water-diversion level known as the "sustainable diversions limit". A Commonwealth government–appointed body would thereafter determine the levels of diversions. A focus on "sustainable diversions" represented a discursive turn towards a focus on the health of the river. The discourse also positioned the Murray–Darling Basin Authority, an appointed body, as having primary control over water. Section 10 of the *Water Act* sets out the basis for Commonwealth management, citing the physical interconnectedness of water resources, the national significance of water in the basin as a resource, the scarcity and further depletion of the resource, the environmental significance and the potential for detrimental economic and social impacts on the wellbeing of the communities in the Murray–Darling Basin. As the Commonwealth government began to intervene in states' decisions more directly (water management had always been under the constitutional purview of the states), the governance discourse shifted towards "sustainable diversion limits" and "environmental water".[13] Sustainable diversion limits are the "maximum amount of surface and groundwater that can be taken from the basin for agricultural and human consumptive use".[14] The Murray–Darling Basin Authority estimated that the environmentally sustainable take is 10,873 gigalitres per year, averaged over the long term.[15] Water beyond the sustainable diversion limit, allocated and managed explicitly to improve the health of rivers, wetlands and floodplains, is known as "environmental water". These terms created

---

13 As of 2016, the federal government controls a majority of the water in the basin. But water diversions are administered by numerous levels of government and other bureaucratic entities. Helen Dalton told me that for every farming business in the basin there is at least one bureaucrat managing water (H. Dalton, personal communication, 2016). Just in the southern connected system, Dalton calculated that there are 17 different bureaucracies including the Commonwealth, the states, Snowy Hydro, National Water Commission, WaterNSW, Murrumbidgee Irrigation, Coleambally Irrigation, Office of Environment and Heritage, the Department of Forestry, the Bureau of Meteorology and many more (H. Dalton, personal communication, 2016).

14 Australian Government Department of Agriculture, Water and the Environment 2019.

15 CSIRO 2020.

a clear distinction regarding the classifications of "environmental" and "productive" water. I will come back to these terms later.

The Commonwealth *Water Act* of 2007, administered under the oversight of the Murray–Darling Basin Authority, includes three central tenets: first, to divert water through administrative and legislative changes; second, to create public investment in water irrigation infrastructure meant to create water savings; third, to use government funds to buy back water to tackle over-allocation. "Water buybacks", though represented as a last resort, gained much attention after the two other approaches largely failed to provide results.[16] The political costs of meeting the sustainable diversion limits through administrative and legislative changes would prove to be too costly, so very little action of this kind was taken. As will be discussed in the next section on economic rationalism, legislative enforcement without a distinct monetary mechanism for reallocating water would not have been well received by the farming community. The second measure, improving infrastructure through public funding, also proved to be largely ineffective. This is because infrastructure is subject to diminishing returns and the water yields from such initiatives are quite modest.[17]

According to the Murray–Darling Basin Authority's plan to implement the provisions of the Act, several measures were to be implemented, including the recovery of an average of 2,750 gigalitres per year from consumptive use towards environmental flows, mainly through buybacks. The plan involved a multibillion-dollar investment in irrigational infrastructure modernisation and the purchase of water entitlements.[18]

It is important to note that this plan may have been too ambitious, as the Commonwealth could not carry it through. For example, determining the "optimal amounts" of water diversions for environmental purposes is an imperfect science. Calculating "optimal" water diversions involves balancing the trade-offs between the net benefits of allocating water for irrigated agriculture and other purposes versus the costs of reduced surface flows for the environment.

---

16   Crase, O'Keefe and Kinoshita 2012.
17   Baumgartner, Barlow et al. 2021.
18   Swirepik, Burns et al. 2015.

According to Grafton, Chu and colleagues, water planners do not have the tools to allocate water optimally among competing uses.[19] Other scholars and critics have argued that it is impossible to calculate with any certainty the water requirements of all the ecosystems within the basin.[20] These factors made it difficult for farmers and irrigators to accept the new plan, especially when their livelihoods were undermined by it, particularly as many were still coping with the adjustments made after the cap had been introduced. Farmers saw the ecological problems in the basin as being related to numerous problems and were concerned about the government's primary focus on flows. They also resented that they had spent years negotiating the cap, only to have more water taken out of the system before the effects of the cap could be adequately measured. Farmers wanted to see ample evidence that the government's more centralised approach would bring results, particularly where the sustainable diversion limits undermined their livelihoods. This combination of an unwieldy and overambitious plan, and the criticism it received from farmers, led to a crisis of confidence in the Commonwealth government. I see all this as an unintended productive effect of adherence to a discourse of administrative rationalism that privileged government expertise and a largely top-down management approach, notwithstanding that approach's genuine limitations.

As we saw in Chapter 1, the history of the Murray–Darling Basin was characterised by state-led intervention from early in the 20th century to the present day. When times were good, no one complained about the massive infusion of government money into irrigation agriculture. The irrigation companies that were owned and operated by New South Wales and the Commonwealth set the parameters for water usage. But as the consequences of "over-allocation" gained more attention, Commonwealth and state governments tried to limit their role in enforcing regulations. This approach placed the burdens of the system on the irrigators. Governments were enthusiastic to intervene in conserving the environment as they considered this within their realm of accountability. But, in terms of the economic and social costs of a

---

19   Grafton, Chu et al. 2011.
20   Swirepik, Burns et al. 2015.

system in collapse, shouldering the burden of those costs was largely considered outside the government's mandate.

While I have discussed the productive effects of administrative rationalism in the Murray–Darling Basin, there were also several disciplinary effects. Four specific examples illustrate these effects. The first was that governments tended to focus on the volume of water in the system rather than its quality, sometimes leading to adverse blackwater events and other consequences.[21] Second, governments focused on protecting certain sites by increasing the volume of water but failed to acknowledge some barriers to getting water to those sites. Third, while the government focused on getting water to certain areas, they often did not consider the dramatic effects of flooding en route to the designated sites, sometimes with catastrophic results for farms. Finally, the focus on delivering "environmental water" through the system caused governments to underestimate how some irrigation users were abusing the system, as in Broken Hill. The focus on delivering sufficient flows gave people false confidence in the government's ability to deliver results, a confidence that was shattered by events at Broken Hill (discussed later in this section).

The Commonwealth government's focus on water flows over water quality was concerning for many farmers. The government's emphasis on a target that can be sold to the public and achievable through the government represents a narrow framing of the issue imposed by the administrative rationalist discourse. Other potential actions like improving water quality through revegetation were effectively "disciplined" out of consideration. Some farmers said the government's preoccupation with modelled flow targets for water as the solution to the health of the system was not scientifically justified. As cases of blackwater events revealed, reaching flow targets without dealing with broader ecological needs and the consequences of increased flow rates can have devastating consequences. As such, blackwater events appear to be a consequence of a predominantly top-down approach to the

---

21   Blackwater events occur when there is a lack of oxygen in the water resulting in mass die-offs of fish. This can occur when excess water causes trees and other debris to fill the water, causing the water to become hypoxic.

management of the basin. (Blackwater events will be discussed in detail in Chapter 4 on green environmentalism.)

The second disciplinary effect of administrative rationalism was that Commonwealth and state governments focused on protecting certain sites by increasing the volume of water but failed to acknowledge barriers to getting water to those sites. The Commonwealth government's decision to relax "constraints" along the river system caused significant third-party effects on farmers operating near the rivers. The failure to acknowledge barriers was partly attributable to the disciplinary effects of technocratic reasoning. Government officials believed that what they were doing was "right", based on expert advice compared to local communities' lay knowledge and personal experiences. An impact of the problem definition is thus that third-party effects are downplayed by simply being referred to as "constraints". The NSW Irrigators' Council's chief executive officer, Andrew Gregson, said the government has failed to recognise third-party effects:

> What they call constraints, we call roads, bridges, towns, farms, and houses. This isn't simply a problem that can be solved by throwing money at it. It's absurd to be increasing the volume of water that you're seeking to acquire without a plan to use it or know the implications of trying to deliver it … a failure to address that will likely see the entire basin plan fail.[22]

A third disciplinary effect I identified was that while the Commonwealth and state governments focused on getting water to certain areas, they often overlooked the effects of flooding en route to the designated sites. In 2016, the New South Wales water office sent water down the river from the Burrinjuck Dam during significant rainfall, wiping out numerous farms. This and similar events led to calls from the communities to conduct an urgent review of water regulations, which revealed that more than 30,000 megalitres were released to boost river health during one of the most significant rainfalls in years. President of the Ricegrowers' Association of Australia,

---

22   Condon 2012.

Jeremy Morton, described these actions by the New South Wales water office as uncoordinated and non-transparent. Morton commented:

> following good rains, irrigators have seen massive amounts of water released from storages and are wondering why it has been happening. We've heard the explanation from Water NSW that translucent flows are required under the water-sharing plans, and we understand natural flows need to be replicated to some degree. You have to question the system though ... At any given time, we don't know the breakdown of water leaving the dam – who owns it, and what its intended use is. This information isn't reported in a single place where irrigators can go to find out whose water is flowing down the river.[23]

The fourth disciplinary effect I identified was that the focus on delivering "environmental water" through the system caused governments sometimes to overlook how some irrigation users were abusing the system, as was the case in Broken Hill in 2017. At that time, this community, led by the town's mayor, demanded an immediate inquiry into the operation of the Murray–Darling Basin Authority. Anger erupted after revelations by the Australian Broadcasting Corporation's *Four Corners* television program that the Commonwealth had sold billions of litres of water set aside for the environment to large-scale irrigated cotton farms in northern New South Wales.[24]

Darriea Turley, the mayor of Broken Hill, commented: "[We are] absolutely outraged. We suspected it, but every time we raised it every politician would push back on us." In 2015, Broken Hill residents had started to notice a bad smell from their water, and some reported suffering from skin conditions, because their main supply – from the Darling River flows – had dried up. Joanie Sanders, a resident of Broken Hill, said, "watching the show and seeing the amount of water upstream irrigators had stored when we had no water in town; it was

---

23 Tempers increase as dam drops 2016.
24 Pumped: Who's benefitting from the billions spent on the Murray–Darling? 2017.

heartbreaking". Further downstream, at Jamesville Station, irrigator Alan Whyte said he was not surprised by anything he saw in the program: "There are 70 families living in this area, there are about 50 properties, and about a quarter of a million sheep, they all literally ran out of water."[25] The Commonwealth had always framed the buybacks as an environmental imperative. The residents of Broken Hill had been able to accept their agricultural and household water shortages mainly because they believed that the government was protecting the wider system by strategically using environmental water. For the residents of Broken Hill and all those in the system who had experienced shortages and been subject to buybacks, the government's actions were a slap in the face. These kinds of top-down decisions on the part of the Commonwealth government significantly undermined confidence in its capacity to deliver equitable outcomes.

Each of these examples demonstrates the real-world disciplinary effects of administrative rationalism discourse and how, notwithstanding a drive towards more free market economic policies, it has historically been and remains the dominant discourse in water policy in the Murray–Darling Basin. Consequently, potential approaches like improving water quality through revegetation or building resilience into the system through water savings measures were often overlooked. Further, incidents of corruption or mismanagement were silenced. The discourse enabled the Commonwealth to take a top-down approach to water management planning and expand its role in owning and controlling water. The evidence suggests that the crisis brought on by the drought further centralised water management despite an environment of free market economic reform.

The discourse of administrative rationalism played a role in shaping the kinds of solutions that policy practitioners put forward. The complex and evolving situation in the Murray–Darling Basin made it difficult for the Commonwealth government to adjust regulations to address the ongoing decline in environmental and resource security. Further, the quality of scientific data and monitoring infrastructure often simply did not meet the requirements dictated by the legislation.

---

25 Murray–Darling Basin: Angry communities call for inquiry 2017.

Murray–Darling Basin Authority policies did not appear to consider the tangible constraints of day-to-day implementation problems, partly because they were rooted in a public philosophy favouring high-level, bureaucratic and scientific planning. Successive Commonwealth governments were discouraged from taking on long-term initiatives that did not benefit them in terms of re-election. In this way, governmental action often took on a reactionary tone. While this is also an institutional problem, political parties often seek to discursively undermine the efforts of the party in power to gain electoral favour. An imperative to appear responsive to crisis and demonstrate confidence and authority in their responses generates an expedient and top-down approach to policy formation.

Sometimes it may be necessary for the government to take a central leadership role, as with the National Water Initiative, but top-down policy prescriptions fail to meet the needs of both the environment and the economy. The discourse of administrative rationalism is pervasive and often limits a more holistic understanding of the system's needs, hindering the ability to mobilise efficient and long-term solutions to the basin's problems. The policy changes since the 1980s, particularly since the Millennium Drought, demonstrate that state and Commonwealth policy is influenced considerably by policy definitions stemming from administrative rationalism. Evidence suggests that the long history of governmental intervention in agricultural modernisation did not end with market-based instruments but rather was an extension of the old paradigm, the major difference being that farmers would bear the costs of the new reforms. As we will see in the following section, the discourse of economic rationalism was, in considerable measure, still influenced by the overarching ideas of administrative rationalism as summarised in Table 3.1.

Economic rationalism and the drive towards free-market reform

After the 1994 Council of Australian Governments water reform, the states separated the land from water to enable "market-based" mechanisms for the water trade. This change meant farmers and other major water users could trade temporary and permanent water

Table 3.1 Administrative rationalism: transcripts and policy, legislation and actions in the Murray–Darling Basin

Transcripts, metaphors and rhetorical devices	Policy, legislation and actions taken
pragmatic	Goulburn Weir (1891)
accountable	Hume Dam (1936)
oversight	Burrendong Dam (1967)
science-based decision making	Snowy Mountain Hydroelectric Scheme (1974)
experts	
efficiencies	National Water Initiative (2004)
sustainable diversion limits	*Water Act* (2007)
modernisation	
constraints	
cap	
water scarcity	
over-allocation	
environmental buybacks	

independent of their land. Under the new water reform rules, water and land became separate titles, which meant that water could be moved easily and flexibly. Increased demand for water was one of the results of this move. Since the mid- to late 1990s, while droughts have affected supply, overall demand has increased. The value of water went up, not only because of the Commonwealth government's large-scale water purchases for environmental flows but also because of the flexibility built into the new system that allowed water to be bought more easily.

The separation of water from land has created many new opportunities for farmers and has, in several cases, allowed the free market to operate more efficiently. Many of the policy changes in the Murray–Darling Basin since the Millennium Drought have been geared towards greater certainty and flexibility in the water market. But creating a market for water as a tradeable commodity has created numerous challenges. Market speculation can pressure farmers to use all their available water, artificially inflating prices and contributing to a focus on economic as opposed to social values. The separation of land from the water had been a productive effect of the discourse of economic rationalism since the 1980s.[26] It changed farmers' understanding of what they could grow and how they could grow it. As discussed in Chapter 1, the Murray–Darling Basin Agreement in 1992, followed by a cap on water diversions in 1996, changed the nature of farming in the basin. The National Water Initiative of 2004 and the *Water Act* (2007) solidified these initiatives. But in many cases, new solutions increased risk and potential failures and spurred environmental buybacks instead of adjustment programs.

While the discourse of administrative rationalism stresses the role of the government, economic rationalism stresses the role of the individual. It reinforces the notion that decision-making should happen at the individual level to promote competition and positive economic growth. Economic rationalism also emphasises that free markets are the best method of decentralising environmental planning. In Australia, at least on the surface, the use of a water-buyback policy for environmental water is in line with the discourse of economic rationalism.

This section outlines the significant disciplinary and productive effects of framing water management in the Murray–Darling Basin in market terms. These include the productive effect of inflation on water prices through speculation. This section also discusses disciplinary effects associated with a focus on economic values, notably reduced concern for social values and a rejection of government interventions that might resemble "welfare". This section also highlights the tensions between market-based approaches and ongoing government intervention. Increased government intervention is thus another critical productive effect of adopting market-based solutions in this case.

As governments adopted the discourse of economic rationalism, farmers increasingly framed their critiques this way as well. Farmers resisted through alternative readings of key concepts like "efficiency" and "productivity". They sought to subvert government policy by challenging and reframing specific terms like "entitlement", "high value" and "risk".

---

26 Crase, O'Keefe and Kinoshita 2012.

The swing towards economic rationalist discourse on the part of government and farmers alike meant some other potentially important ideas and approaches to managing the basin were effectively pushed off the table. For example, the possibility of providing government funding to help defer payments for irrigation services during the drought was marginalised, even by many farmers, to their own detriment.

The remainder of this section focuses on the productive and disciplinary effects of specific transcripts, metaphors and rhetorical devices associated with economic rationalism, including "entitlement", "high-value", "productivity", "efficiency" and "stranded assets". These transcripts are emblematic of the discourse of economic rationalism and have significant productive and disciplinary effects on water management policy in the Murray–Darling Basin. "Efficiency" was reframed by farmers – albeit still within the discourse of economic rationalism – and they used "stranded assets" to challenge the government's discourse of economic rationalism.

Entitlements are the right to a certain percentage of available water under certain circumstances, with the possibility that the entitlement can be lost under certain conditions. Farmers are issued a percentage of their entitlement every year depending on water availability, which could be nothing in the case of drought. The Murray–Darling Basin Authority described entitlements as "the ongoing right to a share of the available water in the river system up to a maximum amount". This is not to be confused with an allocation. For example, a farmer might own an entitlement that gives them the right to a maximum of 100 megalitres of water each year, although they are not guaranteed to receive the entire 100 megalitres of water in a particular year. The amount they get depends on overall water availability. For example, in a dry year, the farmer might only receive a 50 per cent allocation of their entitlement, or 50 megalitres. Entitlements can be bought or sold, but once sold, the seller loses the right to a regular share of the water.[27] Farmers are also able to sell their entitlement on the open market. Part of the National Water Initiative plan included a strategy to make water trading easier by separating water entitlements from regulations that define when and under what conditions the entitlements can be used.

---

27   Murray–Darling Basin Authority 2020b.

This move was meant to make it easier to trade from one hydrological system to another and represents the adoption of a more free-market approach. This new policy would create a water entitlement right to a proportion of a consumptive pool.

This change to how entitlements were framed in the National Water Initiative was deeply controversial since it had significant implications for farmers. First, the approach assumes that the regulatory approvals in different areas of the pool are similar enough so that they will not create huge variations in the value of water entitlements. But, as water markets mature, the regulatory approvals of a given area will likely be reflected in the price of the water entitlement. For instance, a water entitlement that can only be used in a limited number of circumstances will be less valuable than one that can be used in many. Second, various conditions influence the value of entitlements, thus calling into question the use of the term "entitlement". For instance, historically, if an entitlement holder did not use their entitlement, they could still retain it. With the policy change, farmers lose their entitlement if they do not use it. Third, farmers pay for entitlements yearly but are not always issued water, since this depends on availability. The entitlements can thus be worth less than nothing. Finally, in cases of limited water availability, the farmer must still pay for all the irrigation infrastructure that is not being used. The fact that a farmer bears the full cost of the system, whether they benefit from it or not, leads them to question what exactly they are entitled to.

From the Commonwealth government's perspective, operating within the discourse of economic rationalism, the only genuine concern is whether the farmer will be compensated if they lose their entitlement. For one government official, irrigators have gained from water reform.[28] The official commented: "as soon as that was converted to a lasting entitlement and as soon as water policy changed around the country, that gave them some real security of tenure, and in theory, at least they can be compensated if you take it away". The rationale behind this approach is that innovative farmers will continue farming while others will be compensated for exiting the industry. The official continued:

---

28   Anonymous, personal communication, 2016.

now they have a transferable, bankable product. Many farmers grew in wealth by millions of dollars overnight almost. They have many more options under the water market than they had before, there is much more rapid investment than there ever was before. Basically, they are great at whingeing, they really like to focus on all the bad things that they think are happening, and they just ignore the good things, it's just appalling.

If we consider only the financial compensation, farmers have been treated fairly. Nevertheless, if we consider that for most farmers, farming is not just a business but a way of life, they have something to "whinge" about. Farm families have often retained and worked the same land for over a century but are then asked to abandon their farms. In most cases, farmers are being asked to give up the only work they have ever known, the work that defines who they are as people and their vital role within the larger community.

The farmer does not view a water entitlement as a simple monetary instrument: the entitlement can represent something much more valuable. While it can be argued that they can simply sell their entitlement and buy temporary water, the farmer is giving up what is, by definition, an implied security. This is a poignant example of how a term can suggest one meaning but, in practice, mean something entirely different, with far-reaching productive effects. In this case, the productive effect of this new way of defining entitlements is that, for some farmers, their water entitlements have either lost their value entirely or at least have the potential to do so under the right circumstances. These farmers can no longer depend on their entitlements to make decisions, which can undermine the viability of their businesses and sometimes put them out of business.

Governments have endeavoured to frame entitlements within the language of the market but framing entitlements in this new way raises important questions about infrastructure. The value of entitlement is closely related to the infrastructure surrounding it. The costs of maintaining a farming system made possible primarily through irrigation infrastructure projects raises the question of who should pay for infrastructure. Further, infrastructure determines the worth of the various entitlements distributed within the system. As was discussed in

Chapter 1, the irrigation system in Australia was historically designed and built through government-led projects. The New South Wales Irrigation Council and irrigators' groups in the other states are responsible for the upkeep of all established irrigation infrastructure. Farmers said the irrigation community has borne the costs of drought as much as possible. In contrast, some people in the Commonwealth believe that the government has done too much to prop up a failing industry by supporting infrastructure projects.[29] One productive effect of the term "entitlements" is a dilemma of responsibility. If the infrastructure was developed by governments and upheld by farmers, this calls into question who is responsible for the long-term upkeep of infrastructure.

Herein lies a counter-reading of entitlement, wherein entitlement also implies a responsibility to pay for infrastructure. The way entitlements are currently framed has the disciplinary effect of placing responsibility on farmers and implicitly denying responsibility on the part of the government. Entitlements defined by the Murray–Darling Basin Authority are based on water availability, but the farmer pays the price regardless of the reasons for reduced availability. For example, when infrastructure is dismantled due to government buybacks, the new entitlement holder (that is, the government) does not bear the responsibility for the costs associated with these assets disappearing. This represents a disciplinary effect of the language of entitlement as it appears in the Murray–Darling Basin.

One government official alleged: "they [farmers] are so used to governments coming in and giving them infrastructure grants for nothing. No other industry really gets that sort of largesse".[30] As evidenced in passages like the one below, the Commonwealth government understands that investing in agricultural infrastructure projects implies support for those industries in the long run:

> even now there is a big push to develop more water infrastructure, especially in Northern Australia. The big worry is that you pay for the infrastructure, and then the farmers go there, and they

---

29 Anonymous, personal communication, 2016.
30 Anonymous, personal communication, 2016.

can't afford to pay for the upkeep, and the farms fail, especially in the north-west where there are much more variable systems, the soils are crap, and there is nowhere near the dams anyway. We are talking about projects that are way more likely to have a benefit–cost ratio that is not good, but still, there are calls for government money to go to such projects. When you are dealing with public money, I think it is really important to firstly consider what your infrastructure priorities are across government, and secondly, what's going to deliver the best economic return, and finally what kind of environmental impact is it going to have.[31]

The public gained economically from such projects when infrastructure was developed in the basin. This history suggests that the government would have an ongoing role in protecting these investments for the public good. Instead, the discourse of economic rationalism focuses on the value of entitlements based on market conditions. But the quality and strategic placement of water infrastructure is perhaps the most significant determinant of the value of an entitlement. The discourse has the disciplinary effect of downplaying or dismissing an ongoing role in governmental investments. Further, water reform has, over time, eroded the previous meaning of entitlements because their value is undermined by unpredictable and increasingly lower allocations on the part of the government. The resulting policy solutions are thus seen to have little value in the eyes of farmers. Given the unpredictability of the worth of their current entitlements, it is difficult for farmers to trust that the national government will not take them away or that they will retain any worth under entitlement restrictions.

This example demonstrates how governments seek to define a problem under the umbrella of economic rationalism but that the legacy of administrative rationalism continues to be pervasive when it comes to dictating the conditions of water policy in the basin. It reinforces the arguments previously made around administrative rationalism's pervasive effects on economic rationalism's discourse.

The term "high value" was also critical to the discourse of economic rationalism. "High value" is generally used to refer to specific crops that

---

31   Anonymous, personal communication, 2016.

can be sold at higher prices or to refer to the users who grow them: "high-value users".[32] The term "high value" has far-reaching productive and disciplinary effects in the Murray–Darling Basin. It is synonymous with the turn towards a market-oriented system and economic rationalism generally. A productive effect of its use is to put increased focus on extracting optimal value from water and an impetus towards using every drop of water. A focus on "value", even though it is a market-based term, increases the need for government interventions to curb the over-use of water. A focus on "value" also leads to permanent plantings being favoured over annual ones, with significant long-term consequences. Further, emphasising high-value crops can increase market speculation and inflate market prices. The term "high value" also has disciplinary effects, such as attention to monetary aspects of value and overlooking other types of values. In addition, it ignores the effects of moving water out of regions more suitable for "lower-value" crops. Consequently, a further productive effect of thinking in terms of "value" is to move large amounts of water out of watersheds that are more suited to growing lower-value crops, with potentially devastating consequences for the health and sustainability of the system.

While the restrictive water trading of the past was thought to result in lost economic opportunities and environmental damage, some research shows that historically a significant proportion of diverted water returned to the rivers and streams.[33] The development of water markets made water a valuable trading commodity, increasing the volume diverted for production. The diversions go towards higher-value crops, meaning less water is returned to the environment. The language of scarcity is synonymous with "value". An approach that seeks to ensure that every drop of water is accounted for creates a system in which water becomes so valuable as a trading commodity that crops use an ever-increasing percentage of diverted water. In some sense, historically, at least, "wasted" water had not been wasted at all. A focus on the "value" of water puts pressure on farmers to use their water, even in drought conditions. After the *Water Act* was implemented, higher-value crops began to be grown, particularly almonds. This involved

---

32   Murray–Darling Basin Authority 2021c, 2020a.
33   Connell 2007.

investment schemes by large companies that put in big plantings. They bought up a lot of the "sleeper" water licences (which use none – or little – of their allocation or entitlement over the course of a year but have the potential to be used in the future) and quickly activated them.[34] Consequently, the system had much less water when the drought began, because the allocations were being used fully.[35]

In the past, farmers could keep producing even during severe drought because the water allocations were not fully used. One farmer, Ian Mason, recalled how earlier droughts, particularly in the 1960s, were often a blessing for irrigation farmers:

> Quite a perverse outcome in many ways, in those days, irrigation farmers, because dams were never drawn right down, they were able to keep producing. So, they often made money during a drought because they could produce when the dryland farmers couldn't. I know that sounds counter-intuitive, but that was what used to happen. That can't happen now because it's not like that anymore.[36]

Even during a drought, there was still enough water in the system for irrigators to run a typical farming program.

Economic rationalism is based on the premise that markets will determine the best possible use of a resource. From the government's perspective, it is its responsibility to ensure that regulations allow the free market to operate. The unfettered market, in theory, delivers water to the highest-value users. In other words, the user with the highest likely return on investment and available funds buys the water. As Russell James of the Murray–Darling Basin explained:

> in a sort of modern economy, if our goal is to maximise the utility of that water that is available for extraction and use, then the best thing the government can do is establish water access arrangement for people who want to use the water ... You

---

34 Davies 2018.
35 D. Schoen, personal communication, 2016.
36 I. Mason, personal communication, 2016.

establish a market and in theory, over time, the people who value the water the most will be the ones that hold those licences and use the water.

According to James, the water trading system allowed permanent plantings to stay alive during the drought. For almost three years, virtually no annual crops like rice or cotton were grown. In many cases, the permanent plantings were also unable to produce much, but the trading regime allowed those permanent plantings to survive. Further, the people who would have normally grown rice could still retain some income by selling their annual licences to the growers with permanent plantings. For James, the result was that "the people who produced nothing still got an income, and the people who valued their crops more highly were able to protect those crops".[37]

Some farmers shared James's sentiment, because the benefits of water trading were substantial for them, given that their water entitlements exceeded their irrigation needs. Farmer Bruce Atkinson has a big entitlement, but much of his farm is not suitable for irrigation because it is undulating, which makes for very inefficient irrigation. For Atkinson, quite often, it is more economical to sell the water than to use it: "[T]here are new industries that can pay a lot more for water like raisins and almonds, other horticultural products, and cotton. We are better off selling our water to someone who can pay more for water."[38] Very little rice was grown in that period because growers could not afford the water. Therefore, water went to permanent plantings that could absorb the extra costs.[39]

Despite the benefits of moving towards growing high-value crops and for others to sell their water, there were also significant problems. For instance, a productive effect of this way of constructing the value of a crop is that it can generate a landscape with less resilience or flexibility. Defining a high-value crop is tricky, because if everyone starts growing it, it becomes surplus and eventually becomes a low-value crop. Farmer Hayden Cudmore, for instance, believes that the

---

37   R. James, personal communication, 2016.
38   B. Atkinson, personal communication, 2016.
39   R. Sagwood, personal communication, 2016.

transition towards more high-value crops is generally good but is concerned that the long-term implications of such a transition have not been fully considered. High-value crops like almonds are generally permanent plantings that entail a whole different set of risks and responsibilities:

> I think that what we have always done is good but over time, there needs to be some permanent plantings that are your higher value, but we can't all grow the same because we don't want to create an oversupply, and second, the water situation, in terms of availability would not be able to cope with enough water supply to maintain all permanent plantings. The way the licences work, we have some water that is higher reliability than others, and what I do in annual cropping and rice production is an annual crop, so if the water is available, I grow it, and if the water's not, then I can opt-out. Whereas, with permanent plantings, you can't opt-in and out.

Transitioning to permanent plantings means a longer-term commitment to irrigation. Permanent plantings are not just high value; they also require a higher investment and incur higher risk if farmers do not have sufficient water rights. During the drought years, Cudmore did not experience crop losses because he simply chose not to grow a rice crop. He remarked: "the crops didn't fail in those dry years; I was just not able to get them in the ground in the first place".[40]

To complicate these issues, farmers are also disproportionally susceptible to market conditions that include the entrance of new investors. Thus, another potential productive effect of focusing on "high-value" crops is increasing financialisation. Investors and speculators with deep pockets can artificially drive up the price of water and make farming financially unviable. Financialisation has made farmers uneasy because water could go not to the "highest-value" crops but simply to the highest investor. Trading water could drive up its price while reducing productive use simultaneously. One farmer, Bruce Atkinson, explained this concern:

---

40 H. Cudmore, personal communication, 2016.

There is only one way to get water back, and that is through the market and it will be an open market system because it can't be anything else. People with deep pockets will buy it, but they won't be from here; they will be from down the river, it will be an institution or a superannuation company, or some sort of corporate will buy the water. If the Commonwealth did sell the water back to industry, it won't come back to where it came from here.[41]

Another effect of separating water from the land was the tendency for water to change hands solely to increase profits for investors. For example, many people had high debts during and after the drought, so they sold off water to pay off those debts. The problem was that it was not farmers buying the water. One farmer, Allen Clark, explained the problem from his perspective:

Trading has its advantages, but when third parties have the water and not farmers, and when investors get involved, then they try to find ways to push the price of water up artificially, which can really hurt farmers. If an investor buys water for 1,500 dollars, then he is expecting a return of 6 or 7 per cent. As everyone is chasing water and buying water up, that price might increase to 3,000 dollars, so investors can drive the price way up.[42]

Financialisation of water can lead to artificial increases in prices due to speculation in the market. This problem is also closely associated with the difficulties of managing risk. When farmers are uncertain about their future allocations for the next season (and for their livestock), they tend to buy more water on the temporary market, pushing the water price up even if there is plenty of water in the system. Allen Clark observed that even when the weather is just slightly dry, there is great concern that the price will increase, so everyone rushes to buy water before the price increases, which essentially causes the price of

---

41 B. Atkinson, personal communication, 2016.
42 A. Clark, personal communication, 2016.

water to increase. Insecurity in the market pushes the price of water up artificially.[43]

There was a consensus among the farmers I spoke with that speculation has contributed to driving up water prices, but economic rationalism reinforces the view that such financial instruments are simply part of doing business. Economic rationalism also reinforces the notion that despite market conditions, it is the responsibility of farmers to respond to the market. For some farmers like Cudmore, skilled at buying and selling water, such a system can offer benefits. For others like John Bonetti, farming should not be about managing water markets but about the business of farming itself.[44] There are significant issues with treating water like any other commodity that can be bought and sold and is susceptible to speculation.

While there are distinct benefits to water trading, evidence suggests that governments have a role in ensuring fair market competition. Some farmers I spoke with argued that it is also essential to understand that there is a slew of potential values associated with growing crops, not all of which can be easily monetised. Further, in many cases, only the immediate monetary impacts are measured without regard for secondary industries supported by primary production. In sum, the term "high value" is not a simple concept and the policy implications of how we define the concept have substantial productive effects on policy. I now consider some of the disciplinary effects of this concept of "high-value" crops and users associated with economic rationalism in the Murray–Darling Basin.

One significant disciplinary effect of this emphasis on "high value" is that the broader social, ecological and even economic values associated with water use are not considered. Water-value calculations only take into consideration the instrumental value of water among competing users of the water. For example, the approach does not account for the secondary industries that benefit from what happens in Australia or elsewhere. Consider this case. In 2015, the price of water peaked at around $300 a megalitre, which made it very difficult for farmers like Allen Clark to have a good margin (input-to-sale price

---

43   A. Clark, personal communication, 2016.
44   J. Bonetti, personal communication, 2016.

ratio) on any of the crops he traditionally grew. As such, Clark decided to experiment with peppermint, a high-value crop. But he questioned the ethical value of producing peppermint when rice could provide a far more significant benefit to the broader population. Clark remarked: "I look at it differently because I could be growing rice which is needed by 70 per cent of the world's population, but instead, I am growing peppermint for 2 per cent of the world's population."[45] The incentives for high-value crops overlook the considerable challenge of providing food security to the world. For farmers, there is a personal moral obligation to provide nourishing food to people. The concept of a "high-value" use in economic rationalism is determined solely by the monetary value of producing that crop, while social values are set aside.

The way that the term "value" is understood has further implications. For example, farmer Barry Kirkup also challenged the perspective of economic rationalism:

> The way the government looks at it is that it [water] should go to the highest-value users, so whether the food is sent overseas to support value-added industries in other countries, whether it is used in Australia, or whether the crop is being used to produce non-essential items for the middle and upper classes, or to provide food aid to the poor is of no consequence. The government approach is pure economics. There are, therefore, third-party value impacts on water use that are completely overlooked.

Citing a more extreme example, Kirkup worried that there is nothing in the current rules that could prevent someone from simply bottling the water and sending it overseas: "they could get a temporary trade return of 1,000 dollars a megalitre, and they could go and buy it for 150. Whereas, I'm trying to add value at the farm level for food and fibre."[46]

Another critical disciplinary effect of valuing water in economic terms is that it ignores distinct needs and priorities within watersheds by allowing – even encouraging – inter-watershed water trade. Several

---

45   A. Clark, personal communication, 2016.
46   B. Kirkup, personal communication, 2016.

farmers commented that putting a value on the water was a favourable decision. For example, farmer and winemaker Darren De Bortoli commented: "the single best thing the government did was to put a value on the water". But he believed that the significant weakness in this system is that water can be traded between catchment areas. Higher-value crop growers in a completely different region can buy up the water in a catchment and cause that area to become unviable. In these circumstances, whole irrigation areas could become unviable because they are only suitable for lower-value crops like rice. The consequences of moving large volumes of productive water out of an entire catchment area could be devastating. As De Bortoli pointed out: "you can't move water from one part all the way down to another part because someone will pay more somewhere else way down the system. That is a major fault of the current management; the management should be based on catchment areas."[47] It is not always the case that farmers can simply switch to high-value crops. Certain moisture and soil conditions are likely to dictate these types of choices just as much as crop value. It would be impractical to start a farming program of higher-value crops where the soil conditions are not ideal.[48]

In conclusion, the focus on "high value" has the disciplinary effect of overlooking (as irrelevant to determining "value") the environmental and social costs of water trading. Agricultural users are not generally "high-value" users, and it is problematic to assume that a higher-value user (like mining, for example) will be more productive for the economy. There is also widespread fear in local communities that people will sell their water entitlements and exit their communities, leading to a further decrease in rural populations, a reduction in rural services, the closing of local businesses that support farmers and a waning sense of community.[49] Further, marketisation has not significantly reduced the central role of government in practice but has shifted risk and responsibility for negative externalities onto farmers and the communities that rely on them. Despite extensive efforts by the Commonwealth government to reduce interventions in agriculture and

---

47   D. De Bortoli, personal communication, 2016.
48   D. De Bortoli, personal communication, 2016.
49   Kiem 2013, 1623.

let markets perform, the regulations in place appear to do the opposite. In the example of water trading, the government in Canberra does not appear to grasp the full complexity of water as a tradeable asset. The transcript of "high value" in water markets has often unnecessarily increased bureaucratisation and disrupted business operations. This transcript, advanced by the government, has caused confusion and, in some cases, contributed to financial losses and poor environmental outcomes.

Another important transcript associated with economic rationalism in the Murray–Darling Basin is "productivity". An emphasis on "productivity" is illustrated by the creation of the Productivity Commission Inquiry into national water reform under the *Water Act 2007*, which undertakes inquiries into the progress of reform in Australia's water resources sector every three years.[50] "Productivity" is closely linked to the goals of economic rationalism. In the Murray–Darling Basin, it is measured only in terms of output (measured in value) per unit of water used on a farm. It tends to ignore the economic multiplier effects of the value of the activity taking place within the community.

In the past, it was customary for farmers to leave large parts of their farm fallow, but today it is simply impossible to compete if land is left fallow. Farmer Ian Mason recalled that in the past, "the intensity in farming was lower … Nowadays, we farm every hectare. The level of intensity is high. I think there was a natural ability to survive because the intensity wasn't as high, and droughts were shorter."[51] In today's business climate, there are enormous pressures to be productive. Like efficiency, how we define productivity also significantly influences policy choices.

Discussions around productivity have historically focused on getting the most out of the water and maximising the production of a given commodity on the farm. A disciplinary effect of this view of productivity, however, is that it limits the number of factors (or variables) we consider when we determine what being "productive" means in practice. The dominant conception of productivity is limited

---

50   Murray–Darling Basin Authority 2021a.
51   I. Mason, personal communication, 2016.

to the farm, its crops and water use. It does not include productivity in secondary industries that arise out of certain types of production, and it does not consider how many jobs those other industries can create.

From the government's perspective, as articulated by Russell James, it is "not the government's role to pick winners or losers". But in the case of rice growers, a robust secondary industry has grown around rice production. SunRice produces a wide range of rice products and has well-established manufacturing sites employing many people. It has invested significant funds in these sites and depends on a certain amount of rice production yearly to maintain operations.[52] If more water recovery continues to occur, less rice will be grown. As other commodities enter the region, it will be more difficult for SunRice to maintain its operations. Therefore, how productivity is understood is critical. In particular, the question of who is productive and at what scale productivity is measured becomes important.

A related disciplinary effect of the common understanding of productivity is that it excludes consideration of secondary values, like the generation of employment and wealth within the Australian economy. For rice growers, there is increasing competition for water from nuts, mainly cashews and almonds. Nut growers are establishing larger farms in the southern basin, particularly because permanent plantings have a more secure water allocation. Much investment comes from companies based in the United States that buy up water and send the nuts directly to their processing centres in the United States. These exports mean that processing will not happen in Australia, and the secondary value associated with the water used for production will not be retained within the Australian economy.[53] As Russell James explained: "the nuts are basically taken off the tree, de-husked, or whatever, and then put in a packet and sold overseas. That's a fairly light touch in terms of the processing sector." This industry, therefore, has very little productive value to the Australian economy. Although the rice industry has taken the greatest hit when it comes to the free market for water and the impact of drought, it retains one of the highest values

---

52 R. James, personal communication, 2016.
53 R. James, personal communication, 2016.

for the Australian economy in terms of generating income and jobs in secondary industries.

One government official said that, in terms of the rice industry, the secondary processing industry could be maintained by buying rice from elsewhere if necessary: "Even in the rice industry, they found that even when production was way down, they simply bought rice from Pakistan and put it through their production facilities and made the products that they make anyway." The official said that while some of the facilities were forced to shut down, that is just an impact of the drought: "That's inevitable, that's farming in this country, we have such a variable climate; whether it is irrigated agriculture or not we are always vulnerable to drought."[54] These comments highlight the government's limited vision regarding the productive capacity that rice delivers to the broader economy and the community.

The way that productivity is defined within the discourse of economic rationalism narrows the types of policy measures considered. Farming contributes to many secondary productive industries and employment, but the government's productivity measures do not adequately account for these secondary measures. Defining what it means to be productive in a modern economy and balancing production needs with environmental outcomes deserve careful attention. As will be discussed in Chapter 5, farmers often characterise productivity as part of a community-wide issue. For the farmers I spoke with, small businesses and large production facilities like SunRice should all be included in measures of productivity.

Another key transcript observed in the Murray–Darling Basin and emblematic of the discourse of economic rationalism is "efficiency". The farmers I interviewed seek to challenge the dominant framing of "efficiency". Governments have implemented efficiency programs to save water, not increase productivity. On the other hand, farmers wish to increase productivity through the efficiencies they achieve using government programs. Efficiencies gained through upgrades are seen as a way to generate income and grow their businesses. In other words, the farmers are advancing a market-based definition of efficiency that focuses on growth. A key difference is that governments view efficiency

---

54   Anonymous government official, personal communication, 2016.

at the farm level. On the other hand, farmers see efficiency as a value across the entire farming system, not just in reducing water use. They consider increases in electricity, fuel and other inputs when modifying their systems. Farmers seek to challenge the government definition of efficiency by offering an alternative reading focusing on market-based growth and community-wide values. Further, governments have tended to focus on short-term water savings while farmers have asked about the long-term efficiency costs. Farmers seek to redefine "efficiency" within the logic of economic rationalism. Both groups, however, tend to overlook definitions of efficiency that could be considered outside the logic of economic rationalism.

The rationale behind water buybacks in exchange for money towards on-farm efficiency programs is that farmers can reduce their water use and grow more crops with less. Efficiency programs provide government funding for on-farm irrigation systems and on-farm water use reduction measures in exchange for water for environmental purposes. These programs follow a market-based logic as the government exchanges project money for water. Efficiencies include installing new or upgrading existing irrigation infrastructure or technology, including automated water management systems and sensing equipment to improve irrigation efficiency; improving irrigated area layout or design to improve on-farm irrigation efficiency (for example, laser grading or decommissioning old irrigation infrastructure); upgrading, or converting to, surface or sub-surface drip systems and overhead spray systems; and, finally, installing ancillary equipment necessary for new or upgraded irrigation systems to function (for example, computer equipment or pumping equipment). For the government, these "efficiency" measures aim to reduce the amount of water being used on farms and put that water towards the environment. The government is asking the farmers to produce the same amount with less. From the farmers' perspective, being more efficient may reduce water use, but it also increases the potential for productivity. Farmers who develop expensive infrastructure to become more efficient often feel compelled to produce more; farmers, therefore, frame "efficiency" in market-based terms to contest the government's assumptions.

Russell James, a government official with the Murray–Darling Basin Authority, acknowledges that the effect of the plan takes water out of production, which will inevitably affect the level of production: "even though some systems will be more efficient, it will not counter the overall impact, being a reduction in production". According to James, investing in efficiency simply balances the effect of buying back licences and taking water out of production. James commented: "I guess the point is, production shouldn't be too much lower, it will be somewhat lower, but it shouldn't be too much lower and, in the future, particularly when we're moving into a drier climate, the production systems that are left after all of this investment takes place will be more efficient and, in a sense, better positioned to produce in a drying climate."[55]

In contrast to this view, for most farmers I spoke with, the goal of efficiency is to produce more. Farmers cannot understand why they should invest so much time and money into infrastructure if they cannot produce more. After the program is complete, farmers have a new state-of-the-art system, but they still have the same or less water to work with. Allen Clark explained the problem in this way: "The theory of it was quite good, that if I used 500 megalitres of water to irrigate 500 hectares traditionally if it came back that I only needed 400 megalitres to irrigate 500 hectares, and the government bought 100 megs off of the farmer, then that's great. The only problem is that now you have this amazing updated irrigation system, and you wish you could do more with it."[56]

While the Commonwealth government has looked primarily at water savings on individual farms, farmers have asked how water savings in one farm or area can affect efficiencies throughout the larger farm system and dependent communities. Farmers have resisted the way the government defines efficiency by questioning their focus at the farm level. According to farmer John Hand, a buyback might be a win for the individual farmer but is ultimately a loss for the larger community because water is effectively taken out of the community and production is lost: "It's like a milking cow, you have lost the milking

---

55 R. James, personal communication, 2016.
56 A. Clark, personal communication, 2016.

cow. In good faith they thought they would get one and a half times as much from the cow by changing the way we do things, but it's not reflective." Hand provided an analogy explaining how efficiency is understood and operates in reality. I will call it the "efficiency trap".

> Assume you've got 800 cows in a dairy. It is very slow and labour-intensive. So, the government has come and said we will take 300 of those cows, and you can get a rotary dairy. So, you say yeah, because it's going to be a lot more efficient. So, you've given 300 cows to the government, and the government has given you a whack of cash. You build a rotary dairy. So instead of milking 800 cows in two hours, you milk 500 cows in three-quarters of an hour. But the problem is that the thing that gives you the milk is the cow, so now that you are more productive, you think to yourself, I want 1,000 cows because I'm so productive. But now those cows are out of the market. And you have had 4 or 5 people do the same things around you and they are competing for extra cattle as well. And if you can get the cattle on the temporary market you are all chasing it because you are more efficient. Each cow doesn't really give much more milk, only because it's in and out of the dairy quicker, but it's physically restrained by how much it can produce. So, all you have done is make it easier to milk, but you have lost production. It's a false economy.[57]

While slightly tenuous, we can relate the dairy example to water-efficiency programs at a local level. Hand's example shows that, while farmers become more productive through better water-use practices, water sold in a buyback is taken out of the system forever. As more farmers sell their water entitlements, there is much less water in the community as a whole. This situation leaves farmers with more efficient farming systems but a significant shortage of water. The shortage of water drives up the price of water for all the farms. Farmers can no longer afford to pay for water, so their efficient irrigation systems become useless. The government has defined efficiency as an individual value related to reducing water use at the farm level, but this definition

---

57   J. Hand, personal communication, 2016.

does not consider efficiencies in the broader community. The farmers raise the critical point that it is problematic to measure efficiency only at the level of individual outcomes on each farm. The scale at which one measures "efficiency" becomes significant when we examine cases in closer detail.

How the government defines efficiency has had the disciplinary effect of focusing energies only on water-use efficiencies. Farmer solutions, on the other hand, often seek to reduce water usage and energy use in other ways, including fuel, electricity or labour. Hand, for example, questioned the efficiency of transitioning from flood irrigation to overhead irrigation, a central measure taken by the on-farm efficiency programs. He believes that flood irrigation is much more efficient in his location than overhead systems, particularly on a dollar basis. Considering the cost of labour, machinery and fuel, the overhead irrigation system becomes so costly as not to warrant any of the implied efficiencies with the program. As Hand explained, water converts to a specific amount of food no matter what kind of system is in place. From the start to the end of the season, if the farmer has accurate agronomics and their layout is correct, they can accurately predict how much water is needed to convert water to wheat. However, simply putting a sprinkler on the crop will not significantly affect the conversion rate. Hand argued that there is no significant increase in the productive value of water but simply a change in the delivery sequence that makes watering easier for the farmer.[58]

As is demonstrated through these examples, farmers frame efficiency much more broadly than governments. Both the Commonwealth government and farmers are working within a discourse of efficiency. But the government tends to define efficiencies narrowly, which has disciplinary effects in terms of the types of programs and measures it seeks to implement. If we define efficiency simply in terms of water use, we might measure a high level of efficiency. If we include the various other factors involved, we may see a decrease in efficiency. Farmers seek to reframe the way that efficiencies are defined. As such, the contestation of the term "efficiency" can be

---

58   J. Hand, personal communication, 2016.

seen as a site of resistance within the discourse of economic rationalism.

There are also significant disciplinary effects associated with framing efficiency in terms of water savings over the short term (two–three years) and meeting the goals of taking water out of production for environmental purposes. Farmers draw attention to the long-term impacts of reduced efficiencies within the system. Like John Hand, Gary Knagge did not participate in environmental buyback programs because he believed doing so would destroy any chance of a consistent return. In other words, what may be more efficient in the short term becomes far less efficient in the long term. Knagge understood the appeal of selling water back to the government for environmental purposes and being able to improve paddocks with government funds. But the farmer is left with much less water. This would not be a problem if farmers were guaranteed some minimum allocation on their water entitlement, but they may not receive any water at all. He noted: "you can have a perfectly well laid-out farm because of all of the efficiencies that you've done, but if you don't have enough water, then you're not growing anything".[59]

Interestingly, and in line with economic rationalism, some farmers believe that efficiency programs and government assistance represent an affront to fair competition in farming. This signifies that they see it as outside the market-based approach and unacceptable. For farmer John Bonetti: "if you can't compete in the big bad world, get out. Go and do something else. If you can't compete with Usain Bolt, don't get in the race with him." Bonetti understood that free competition in the market tends to result in fewer farmers being able to compete but believed that using government funds to build farm businesses created a situation of dependence.[60]

In the Australian case, efficiency is viewed differently from how it is in other market-based contexts. Under the World Trade Organization Agreement on Agriculture, subsidies are allowed to help farmers to become more efficient without increasing production.[61] These types

59   G. Knagge, personal communication, 2016.
60   J. Bonetti, personal communication, 2016.
61   World Trade Organization 1995.

of programs are generally accepted as part of a market-based system. On-farm efficiency programs in Australia are not considered subsidies according to the World Trade Organization's definition, since they buy water allocations to enable capital improvements. They are market-based, and there is no way for the government to compel farmers to maintain production at the same level or control how farmers spend the money they receive from selling their allocations. This policy means that farmers might increase water use or production because there is no regulatory mechanism for controlling either. Inflating prices for water may not necessarily reduce production and may simply translate into a higher price for the commodity, which is generally passed on to the consumer. In sum, the market-based approach in Australia, with its rejection of subsidies, has significant disciplinary effects on the types of policies that can be implemented.

The government and farmers define "efficiency" within market-based terms and in line with the overarching discourse of economic rationalism. The Commonwealth government focuses on reducing water use through on-farm efficiency programs. Farmers, on the other hand, are focused on increasing production while at the same time reducing water use. Farmers are also more concerned with how water efficiencies affect the broader farming system and farming communities over the long term. Farmers work within the discourse of economic rationalism to resist the government's narrower conception of efficiency. It is notable that both sides overlook alternative definitions of "efficiency" in this context, such as subsidies associated with productivity limits that the World Trade Organization finds acceptable. This is an example of the disciplinary effect of the discourse on all parties. Still, sites of resistance for farmers have emerged within the established discourse of economic rationalism.

The final transcript we will look at is "stranded assets". "Stranded assets" are one of the productive effects of the market-based approach and buybacks. The transcript of "stranded assets" is commonly used within the farming community to refer to abandoned assets or infrastructure as farmers sold their water entitlements. From the perspective of some farmers, the government's market-based approach has shifted financial risk onto farmers and inadvertently undermined their efforts to become more effective environmental stewards. Farmers spend large sums of money to increase their water efficiencies, and even though everyone benefits from such initiatives, only farmers incur the risk from these projects. In the past, governments had invested in dams and other major irrigation projects, but they no longer build the channels and ponds farmers need to manage water effectively. This work is done by the irrigation companies and paid for by farmers. Nonetheless, as farmer Tony Piggins put it, in times of crisis, the governments "come in over the top" and significantly reduce allocations without concern for these investments.[62] Farmers negatively affected by buybacks, and the costs associated with stranded assets, have challenged the economic logic of the government by stating that government interventions essentially undermine the free market.

In the past, banks lent money to farmers, using their water allocation entitlements as security. Large amounts of money were lent to farmers to develop more sustainable irrigation systems. For example, if you had a 5,000-megalitre allocation and the government brought that down to 1,500 megalitres, but you had borrowed against the security of the 5,000, you would be in grave financial trouble.[63] In this situation, the farmer could lose the security on their investment. The farmer may also have to significantly reduce their water usage simultaneously, which leaves any new infrastructure as a "stranded asset". Ultimately, such government intervention can significantly reduce the confidence of both the banks and the farmers to make future investments in infrastructure meant to reduce water usage.

Some farmers expressed the view that they could be doing things more efficiently if the government were more consistent with their way of framing the problem in economic terms. For example, Tony Piggins estimated that he has spent between $500,000 and $750,000 on his irrigation systems. All the pipes and the entire system are underground, so there are no evaporation losses, and they have centre pivots that cost about $100,000 each. Further:

A hundred-acre [40-ha] irrigator is going to cost you in the vicinity of 200,000 dollars. If you run out of water and you can't

---

62   T. Piggins, personal communication, 2016.
63   T. Piggins, personal communication, 2016.

use it, you have all these stranded assets sitting out on the paddocks, and you haven't got the water to put through them, and you can't grow a crop, and no one wants to buy a second-hand irrigator for 200,000 dollars. This is the issue that the government had to confront in this 016 area,[64] which was quite a fraught process. It wasn't all centre-pivot irrigation; there was a lot of flood irrigation happening in 016 as well. There was no way the government could even come close to compensating for the amounts that the farmers had lost.[65]

Not only do farmers experience serious economic hardships, but much of the hard work they have put into water conservation is also undermined, to the detriment of the entire basin system. The financial costs to the farmer and the social costs to the communities that depend on these farms are also far-reaching.

One of the main arguments against the government's approach is that the situation is creating a "Swiss cheese effect" in agriculture with an increasing number of stranded assets (like the holes in Swiss cheese), which makes it harder and costlier to deliver water through the Murray irrigation system (there is less water volume, so the costs of the system are spread over fewer people).

Many of the water licences also have fees associated with them. These fees are designed to underwrite the cost of the infrastructure to deliver water to the farm gate. While this is an onerous financial burden to bear during the drought, some government officials said that the responsibility falls squarely on farmers, and a failure to make these payments results from poor economic planning. One high-level government official, Russell James, stated where he believes the burden of responsibility lies:

> A lot of farmers said, "We can't afford those fees because we're not producing anything." Well, an economic, rational person would say, "Well, that's too bad. You should have provided for that cost because you know it's an annual cost that comes every year. It's the

---

64   These codes are used to identify irrigation zones.
65   T. Piggins, personal communication, 2016.

same cost every year to access those services. You should have put some money away in good times to make up for the fact that in dry years you still have that payment owing."[66]

The costs of buybacks and stranded assets are borne by the farmers who remain in business, creating further incentives for them to exit farming altogether. In addition, the productive use of water in certain areas depends on soil conditions. The piecemeal approach of the Commonwealth government did nothing to ensure that the most productive areas would remain in production. During the drought, some of the most productive farms were in crisis, and farmers struggling with debt were forced to sell. In sum, the logic of economic rationalism has the disciplinary effect of ignoring or discrediting any solutions that are not market-based.

There were several suggestions from individual farmers on how to avoid stranded assets, including not allowing water to be sold between regions so that water remains equally distributed throughout the basin, reducing water allocations equitably based on a percentage of existing allocations of individual farmers, tactically negotiating water acquisitions collectively to minimise losses or to strategically offer water buybacks to minimise the impact of stranded assets in fewer regions. Farmers mobilise this script, "stranded assets", to make these counterpoints to the government. This use of a script is a clear example of farmers working within the discourse of economic rationalism to frame a consequence in terms designed to elicit attention to their concerns. The market-based approach, with its focus on individual compensation, effectively disciplines other approaches and appears to have undermined the cohesion of the broader community by forcing the farmers who remained to shoulder the burden of a reduced water supply.

As evidenced, specific transcripts associated with the discourse of economic rationalism have both productive and disciplinary effects on the kinds of policy options that both government and farmers find acceptable to deal with problems in the Murray–Darling Basin. Both farmers and government officials accept the discourse of economic

---

66   R. James, personal communication, 2016.

rationalism; they tend to value free markets, deregulation and privatisation. But farmers challenge many of the assumptions that are inherent in terms like "entitlement", "high value", "productivity" and "efficiency". Farmers work within the discourse of economic rationalism to challenge ways of framing efficiency and productivity. They also introduce new terms grounded in economic theory to make their case for what they think the governments are missing. The term "stranded assets" is the key example presented here. In this way, farmers challenge and resist elements of policy using the language of economic rationalism. While farmers have mainly accepted the overarching discourse as the reasonable and appropriate response, they have also challenged that discourse.

As explained regarding administrative rationalism, policies characterised by market-based mechanisms tend to hide government interventions that shape the current situation in the Murray–Darling Basin. For instance, while the Commonwealth government has allowed free trade in the water market, it has bought up nearly half that water itself. Further, the market does not determine caps on water and regulations on buyback levels; the government determines them. This means that despite a strong discursive emphasis on economic rationalism, administrative rationalism is still pervasive in practical terms. As was argued in the section on administrative rationalism, there is a long history of government interventions that have a continued effect on policy outcomes in the Murray–Darling Basin. There continue to be numerous governmental rules, regulations and interventions that influence the market in myriad ways. Government investment in infrastructure and involvement in setting caps, buyback levels and efficiency programs all reaffirm its responsibility and involvement in the market.

Farmers resist government interventions from within the discourse of economic rationalism through their alternative readings of entitlement, high value, efficiency and productivity. From within, they offer an alternative definition of efficiency based mainly on the free market philosophy focused on growth. They also resist the discourse from the outside: for example, by reiterating alternative policy responses that were pushed off the table because they did not fit within the market logic, such as the idea that people can only trade within basins and not across them. Both sides miss alternative possibilities, like increasing efficiency while at the same time enforcing regulations that limit production, as is often done by other signatories of the World Trade Organization Agreement on Agriculture. Production limits are used by many countries wherein production levels are maintained at a low level but farmers can increase profits through the best use of water and other inputs.[67] Most significantly, the economic rationalist approach, like the administrative rationalist approach, misses many opportunities because it fails to characterise problems in a way that focuses on social and environmental outcomes. This limitation was evidenced by negative effects on community life and the inadvertent environmental damage like blackwater events that sometimes occurred. (The effects of the current approach to water management will be explored in further detail in Chapters 4 and 5.) Table 3.2 below sums up this section by illustrating the key transcripts, metaphors and rhetorical devices associated with the discourse of economic rationalism, as well as specific pieces of policy, legislation and actions taken in the Murray–Darling Basin that are strongly influenced by this discourse.

Democratic pragmatism

Like administrative rationalism, democratic pragmatism places nature as subordinate to human problem-solving efforts. It is different in that it imagines that problem-solving is a group effort from scientists, elected officials, interest group leaders, voters and non-voters. Everyone has agency, and the democratic process is seen not as hierarchal but as a competition and a cooperative process between actors vying for their interests. Therefore, there is a strong focus on individual agency and autonomy in democratic pragmatism. Independence and self-reliance are thought to bolster personal achievements, which ultimately benefit everyone. In this sense, democratic pragmatism is "a problem-solving discourse reconciled to the basic status quo of liberal capitalism".[68] Individuals have the agency to create collective changes by putting their

---

67   For example in the European Union; Clapp 2016, 71–95.
68   Dryzek 2013, 99.

Table 3.2 Economic rationalism: transcripts and policy, legislation and actions in the Murray–Darling Basin

Transcripts, metaphors and rhetorical devices	Policy, legislation and actions taken
entitlement	National Water Initiative (2004)
high-value user/crop	*Water Act* (2007)
risk	Water trading and the activation of
flexibility	"sleeper" licences
commodities	Buyback and efficiency programs
stranded assets	Financial speculation and
efficiencies	international water markets
productivity	
market-based instruments	
free markets	
managing risk	
growth	
competition	

interests forward. But this focus on individual agency can have the productive effect of perpetuating a zero-sum approach to deliberations that does not factor in how the success of one contributes to the success of all. Neither does it consider how one person's loss contributes to the loss of others. Therefore, the focus on individuals can limit the ability of democratic processes to build collective capacity.[69]

In the case of the Murray–Darling Basin, the prevailing discursive frameworks centred around individual responsibility thwart the implementation of more authentic democratic processes and diminish the capacity of communities to work together to develop collective solutions to problems. As discussed in Chapter 2, democratic pragmatism is a participatory form of democratic engagement. The version of democratic pragmatism practised in the Murray–Darling Basin is constrained by the overarching views embedded in administrative rationalism, particularly individualism. Nonetheless, democratic pragmatism is still crucial in defining what is considered an

---

69   Dryzek 2013.

acceptable deliberative process. Using several examples, I show how the discourse of democratic pragmatism affected the consultation process but how attempts at consultation and engagement were generally inconsequential. I argue that this is because the discourse is limited by the internal disciplines associated with it. These include (but are not limited to) asymmetrical power imbalances, physical disciplines that reduce democratic participation (like the capacity to travel to meetings), knowledge-based disciplines (like the inability to communicate in the kinds of ways that farmers understand) and social disciplines (farmers are part of a different social class).

The examples included in this section show that democratic pragmatism uses the language of engagement but has the productive effect of thwarting more authentic efforts at democratic engagement. This section summarises the Murray–Darling Basin's consultative process and explores the productive effects of democratic pragmatism.

The consultative process in the Murray–Darling Basin appears well aligned with the principles described in the discourse of democratic pragmatism. This is true in so much as town hall meetings, consultations and other forms of engagement took place but, as practised in the Murray–Darling Basin, democratic pragmatism is a narrow and shallow conceptualisation of democratic engagement. The consultation process did not enthusiastically seek out – or depend – on broader input from the affected farm communities. Despite speaking of democratic engagement and participation, decision-making was still orchestrated from the centre. This is because there are limitations associated with democratic pragmatism as a problem-solving discourse.

To understand the role of democratic pragmatism and its effects on policy in the Murray–Darling Basin, it is necessary to explain the bureaucratic procedures and instruments that have accompanied it. In 2007, towards the end of the Millennium Drought, Prime Minister John Howard said that he did not think the states were doing enough to manage water resources. Howard announced the plan for the Commonwealth to step in and recover water for the environment and to set up a new organisation to manage the process, the Murray–Darling Basin Authority. The policy and planning division began working on the National Plan for Water Security.[70] The policy and planning division of the Murray–Darling Basin Authority is concerned with the overall basin, whereas other groups are looking at particular sites, most significantly wetlands. Russell James, executive director of policy and planning, said that they have a "committee-type approach" to the evaluation and try to ensure that the public investment in environmental outcomes is responsibly managed.[71] Six governments (Queensland, New South Wales, Victoria, South Australia, the Australian Capital Territory and the Commonwealth government) are involved in the river system's overall management. Each has a different perspective on how to manage the system.[72] David Dreverman, executive director of river management, Murray–Darling Basin Authority, explained that the Commonwealth government manages the competing demands of the different states. When one state wants more, it generally comes at the cost of the other states. The process of having states agree is complicated by the first River Murray Waters Agreement of 1915, which stipulated that every decision is made by consensus and not by a majority vote. In the Australian Constitution, water remains within the jurisdiction of the states, so "the Commonwealth can influence and lead, and fund, and achieve its leadership through funding, but it cannot direct".[73] But, in developing the Murray–Darling Basin Plan, the states did voluntarily defer some powers to the Commonwealth to allow the basin plan to happen. The Murray–Darling Basin Authority, for example, reviews the progress of the states in meeting their objectives in shared agreements.[74]

The Murray–Darling Basin Authority also has a stakeholder engagement plan that includes consultations with many actors. In this list, Dreverman included the Commonwealth government, which owns the water, the Department of Agriculture and Water Resources and its minister. The Department of the Environment and Energy has significant influence because it controls the sizeable environmental

---

70 Australia. Parliament of Australia and John Howard 2007; R. James, personal communication, 2016.
71 R. James, personal communication, 2016.
72 R. James, personal communication, 2016.
73 D. Dreverman, personal communication, 2016.
74 D. Dreverman, personal communication, 2016.

water portfolio. The Murray–Darling Basin Authority also identified over 100 stakeholder groups, which are listed on its website, including people living in the basin and the broader Australian community, industry, conservation, recreation and community groups, local governments, Indigenous peoples, Basin Community Committee, Basin Officials Committee, state government agencies and departments, Commonwealth government agencies and departments, and scientific, technical and policy, and research communities.[75] Russell James of the Murray–Darling Basin Authority explained that they always have a public consultation process whenever they review or amend the Murray–Darling Basin Authority plan. The Murray–Darling Basin Authority has regular meetings with the National Irrigation Council and the largest irrigator group in the basin, the NSW Irrigators' Council.[76]

As we see from the above, the Commonwealth government often points to democratic processes (such as consultations and outreach) as proof of democratic engagement. It typically fails to acknowledge asymmetrical power imbalances. Democratic pragmatism suggests that where institutional structures (like regulatory bodies and planning committees) are firmly in place, democratic problem-solving should be inclusive and effective, but that was not the case in the Murray–Darling Basin. In Griffith, for instance, community members were so upset with how the plan was delivered to the community that they took the Murray–Darling Basin Authority reform plans and burned them as a public protest. Farmer Barry Kirkup described the plan as a "huge, huge shock", and the reaction against the Murray–Darling Basin Authority plan led to large meetings within the communities of Griffith and Deniliquin. While the drought propelled the government to move forward with the plan, it was proposed right when the drought was ending, and farmers were looking forward to finally having access to water and recovering their businesses. According to Kirkup, the community felt that the government's response to the drought was a "knee-jerk" reaction.[77] In the minds of farmers, the plan felt like some

---

75 Murray–Darling Basin Authority 2009a.
76 R. James, personal communication, 2016.
77 B. Kirkup, personal communication, 2016.

arbitrary punishment because they could not prevent the impacts of drought through previous water reform.

Democratic pragmatism was born from a desire to make administration more flexible and responsive to varied circumstances. However, the main impetus of democratic pragmatism is a desire to secure legitimacy for decisions by involving a broader population.[78] In line with democratic pragmatism, the meetings and institutional structures in place for consultation suggest that there were many opportunities for farmers to be included in the process, so it is important to ask why farmers routinely felt their positions were ignored. Interviews with government officials revealed that their attempts at consultation were not well received, but there was a great deal of doubt as to whether anything could change. While many in the farming community acknowledged attempts by the government to consult, there was also a general feeling that the consultations were more of a political imperative or a legal responsibility than an effort to reform policy collaboratively. Efforts to include this subset of the population can thus be seen as a failure. My research suggests that while some of the consultations were indeed tokenistic, many of the failures in the consultation process came from the disciplinary effects of democratic pragmatism itself. For example, while the Commonwealth government sometimes made sincere efforts, there was no systemic effort to identify and address asymmetrical power imbalances that hinder the democratic process. This is a failure of democratic pragmatism, which assumes equality among citizens.[79] As a result, efforts by the government appeared intentionally tokenistic and dismissive at worst. The following subsections explore these internal disciplines of democratic pragmatism and their effects on the process.

Democratic pragmatism in action

The discourse of democratic pragmatism assumes that when individuals can participate in democratic processes, positive outcomes will follow. But individuals always face barriers, largely unacknowledged in

---

78 Dryzek 2013.
79 Dryzek 2013.

democratic pragmatism. Democratic pragmatism does not recognise the hierarchal structures of power of participants and how these affect their capacity or willingness to participate.

I identify four types of barriers to democratic participation that my research shows were unaddressed in the consultation process. First, there were physical barriers to democratic engagement, which included being unable to attend meetings or travel to metropolitan areas to attend important consultations. Second, there were identified knowledge barriers like an inability to access information and data vital to actively participating in the consultation process. Government officials sometimes denied access to data or presented data to farmers they could not interpret and understand. Farmers did not have the same access to knowledge, or their knowledge was not considered. Third, there were barriers related to social position and the social hierarchies of power that exist in society. These had a disciplinary effect on the capacity of farmers to represent themselves as legitimate authorities. Social hierarchies and public perceptions of farmers sometimes contributed to devaluing their knowledge and input in the consultation process. Last, there were time barriers that affected the democratic process. Political pressures to show progressive change within tight political timeframes limited the capacity for comprehensive democratic participation. The discourse of democratic pragmatism is a limiting view of democratic engagement and participation. Without active attention to the limitations associated with democratic pragmatism, even well-meaning policymakers can be prevented from achieving more authentic, deliberative, democratic outcomes.

First, there were significant physical barriers to democratic engagement in the Murray–Darling Basin. Farmers talked about cases where they were unable to attend meetings in town or to travel to far-away metropolitan areas to attend important consultations. Psychological barriers accompanied these physical ones. Interviews revealed that farmers often felt intimidated by the consultation process. The bureaucracy was intimidating, even without the threat of water being taken away. While some farmers believed it was simply an oversight that government representatives do not usually meet in the farmers' environments, others believed it was a deliberate attempt to avoid consultation, as it is difficult for farmers to travel to Canberra or Sydney.

Some government representatives understood these barriers and, in a few cases, attempted to meet with farmers on their land. A few months before my interview with farmer Debbie Buller, Murray–Darling Basin Authority senior economist Phil Townsend visited Buller's farm. Townsend rode the tractor around the farm with Debbie's husband, Stuart. While in his own space, Stuart could clearly explain his problems to Phil. When farmers are in a meeting room, out of their comfort zone, they can feel intimidated and unable to express their concerns. Buller believed that farmers have been raised with a strong respect for authority, and the fact that the Commonwealth government holds all the power over their access to water makes them intimidated in large meetings.[80]

Similarly, farmer Gary Knagge has had high-profile decision-makers come to his farm. Most visitors told him they had never been on a farm, even though many were making big decisions about funding research for agriculture: "[T]hey brought their shiny asses there, and they had been taught everything in the classroom and had no real grounding of what people in the country are trying to do". He continued: "It's like the person who read the book on how to swim, passed the exam on how to swim, but when they got in the water, they drowned." Knagge said that decision-makers are too much in the theoretical world and don't get mud on their boots enough to see what happens at the grassroots level. He told the story of a woman who came with a group from the city but could not walk across the lawn because she was wearing stilettos.[81]

Reaching out to farmers on their land and demonstrating a knowledge of country living lends legitimacy to politicians and bureaucrats. A willingness to engage within a farmer's space demonstrates a recognition of the power dynamics at play. Spaces affect the quality and level of participation in consultations. Farmers spend most of their lives on their farms, so it is important to recognise how taking them out of their own spaces will have many unintended consequences. We cannot assume that an office or a hotel meeting area is a neutral zone, particularly for people who spend most of their

---

80   D. Buller, personal communication, 2016.
81   G. Knagge, personal communication, 2016.

time outside. Similarly, power is affected by the social dynamics that spaces represent. Physical and accompanying psychological barriers can be a significant impediment to authentic deliberation. (In Chapter 5, I explain how community-centrism discourse proposes overcoming some of these challenges.)

Knowledge-based barriers also represented a significant challenge to authentic participation in the Murray–Darling Basin. One such barrier was a lack of access to information and data vital to actively participating in the consultation process. Another was how data was presented to farmers in ways they could not understand. In other cases, farmers were simply denied access to information. As a result, farmers did not have adequate access to knowledge, and their knowledge was often ignored.

The first version of the basin plan came out in 2010, on the heels of the drought. Most of the data used in the plan was based on computer models developed by the Murray–Darling Basin Authority. But computer models are only as good as the information used to develop them. According to farmer Ian Mason, the irrigators asked several times for the government to provide background about how the Murray–Darling Basin computer models were developed. They were never given this information, leaving them frustrated. They knew what the Murray–Darling Basin Authority's models predicted, but they had no idea how the models were established. The irrigators wanted to understand how the models functioned so they could predict and limit third-party effects, or what Mason calls the unintended consequences of releasing higher water flows.[82] Farmers were also frustrated that their knowledge was not used to help develop the models. They felt the consultation process should have begun at the initial modelling stages.

A lack of effort to bring past knowledge into the plan created much anger among farmers who had worked with the Murray–Darling Basin Commission (the Authority's predecessor) for years. Barry Kirkup explained that farmers were already giving up a percentage of water each year for the environment, reflecting a long-term slow change. Farmers felt there was little acknowledgment of what had been done in the past or the progress made in water reform up to that point. They

---

82   I. Mason, personal communication, 2016.

wanted to see how previous reforms were considered in developing and implementing new reforms, what difference they had made and what further reforms they might be facing. The tight timeframes and lack of acknowledgment of previous reforms made under the guidance of the Murray–Darling Basin Commission generated significant uncertainty and distrust of the Commonwealth government.

The ways that farmers and government representatives value certain knowledge also affect the democratic process. Many farmers said that while they had opportunities to engage, they did not feel their opinions were taken seriously. They felt that decisions had already been made, and the meetings were about listening to the government spokesperson, not about taking what farmers had to say seriously: "[O]ur opinions didn't seem to weigh as highly."[83] From the farmers' perspective, local knowledge was ignored, and science was highly valued in the plan's development, even though the science was largely in its infancy.[84]

Farmer testimonies were corroborated by Professor John Briscoe, a scientist and vocal critic of the plan. Briscoe argued that the framers of the *Water Act* conducted their modelling in a "highly secretive" manner: "The MDBA [Murray–Darling Basin Authority] will run the numbers and the science behind closed doors and then tell you the result. The Murray–Darling Basin Plan process was not, in my view, an aberration which can be pinned entirely on the leadership of the MDBA board and management, but intrinsic to the institutional power concentration that is fundamental to the *Water Act 2007*." He wrote:

> In all of my years of public service, often in very sensitive environments, I had never been subject to such an elaborate "confidentiality" process as that embodied in the preparation of the Guide to the Basin Plan. The logical interpretation was that the spirit of the *Water Act* of 2007 (environment first, science will tell, the Commonwealth government will decide, the people will obey) required such a process. The High-Level Panel told the Chair and CEO of the MDBA that they understood that this

---

83   A. Clark, personal communication, 2016.
84   L. Burge, personal communication, 2016.

was what the Act dictated but that it was the role of senior civil servants to explain that this would not, and could not, work. We were given to believe that there was no appetite for such a message at higher levels in the government in Canberra.[85]

For Briscoe, the Act demonstrated an extraordinary and unusual confidence in the role of science in determining environmental needs. The government appeared to stand behind science rather than acknowledge the uncertain linkages between water use and environmental outcomes.

A significant barrier to consultation was also that the Commonwealth government tended not to take the level of knowledge about water reform within the farming community seriously. It was only after a draft plan was released that submissions were called for, and people were asked to reply with their comments. There were thousands of submissions from councils, community organisations, businesses, irrigation groups and irrigators. Farmer Louise Burge recalled that while there were thousands of submissions, they were treated with the same weight. The irrigators group she represented submitted a lengthy report representing 1,600 to 2,000 irrigators. Burge believed it was given the same value as environmental groups that had a pro forma press-button response.[86] Farmers were also frustrated by government officials, who, they believed, were inadequately informed and could not talk about the issues that were important to farmers. Farmer Helen Dalton said that while the Murray–Darling Basin Authority often sent representatives to Griffith to talk to them, the representatives were not well informed about the situation and could not answer any questions. Dalton related how, at one such meeting, a representative had talked about water projections. Dalton ventured that he must have been working for the Bureau of Meteorology because the presentation centred on the fact that the dams were full and there was a lot of rain. This was not news to the participants since everyone was inundated with floodwaters at the time.[87] When the group started asking questions

---

85   Briscoe 2011, 5.
86   L. Burge, personal communication, 2016.
87   H. Dalton, personal communication, 2016.

about what happens to translucent and transparent flows,[88] the representative could not answer. According to Buller, "representatives simply said they will note the question and get back to us, but they never did".[89] Farmers interpreted the sending of ill-equipped representatives by the government as a tactic to avoid conflict with the community that ultimately undermined the government's stated objective of authentic engagement. Sending representatives ticked the consultative box, but an inability to answer questions or engage with farmers in discussions demonstrated a lack of concern for the outcome of the process.

Farmer John Hand said he would also liked to have seen farmers involved in water reform because it could help the government reduce the cost of implementation and achieve better environmental outcomes. Similarly, farmer John Bradford offered the analogy of someone who wants to catch fish: "if you are going to catch fish, you go and talk to a fisherman". He commented that it makes no sense to pay someone with no experience in fishing to go fishing and have them tell fishermen what bait to use.[90] There is a tendency for governments to overlook their position of authority and power, which can act as a hindrance to acknowledging community-based knowledge. As explained in the discussion of administrative rationalism, the tendency to elevate scientific knowledge over farmer knowledge has consequences for policy reform. This tendency is highlighted in how privileging certain types of knowledge has direct consequences for the democratic process itself. Knowledge-based constraints represented a significant barrier to the consultative process. The examples here demonstrate that the discourse and its practices fail to acknowledge the significant effects of knowledge construction and dissemination on the consultative process and, consequently, on policy development.

---

88   A transparent flow occurs in a regulated river system when inflows are passed through a regulating structure – usually a dam – to enable a near-natural flow pulse into the river system. A translucent flow is similar, however only a portion of the inflow volume is passed.
89   D. Buller, personal communication, 2016.
90   J. Hand, personal communication, 2016; J. Bradford, personal communication, 2016.

The case of the Murray–Darling Basin also illustrates that there were barriers related to social position and the hierarchies of power that exist within society. These barriers had a disciplinary effect on whether farmers were able to represent themselves as legitimate authorities. As the following examples demonstrate, social hierarchies and public perceptions of farmers sometimes contributed to devaluing their knowledge and input in the consultative process. A significant social barrier to democratic engagement stemmed from historical tensions between the farm groups and the Murray–Darling Basin Authority, and its predecessor, the Murray–Darling Basin Commission. Hostilities were exacerbated by several factors, particularly by what farmers saw as a quick change in the Commonwealth government's approach that represented an imminent and existential threat. The relationship between the parties resulted in the government taking an evasive approach, particularly with the community around Griffith. A history of distrust between the government and the locals there contributed to a breakdown in consultations.

A history of tension between local underworld figures and the government complicated the consultation process in Griffith. When the *Water Act* was first introduced, it was necessary to initiate a consultation process. However, as Griffith area-based farmer Helen Dalton put it: "Griffith has a history of knocking people off. Griffith has a mafia here, and there are some very feisty people." New South Wales political candidate Donald Mackay went missing from the car park of a hotel in Griffith on 15 July 1977. Though his body was never found, substantial evidence supports the conclusion that he had been murdered. Dalton said that even though most of it is "puff and wind", government representatives did not want to come to Griffith to talk about water. She told me: "We are not the gentle type (we are, but a lot of people here aren't). There are lots of threats."[91] The reputation of Griffith as a hotbed for organised crime and marijuana cultivation is well known in Australia.[92]

Farmer Barry Kirkup recalls that after being notified of the Murray–Darling Basin Authority plan by the irrigation companies, the

---

91   H. Dalton, personal communication, 2016.
92   Stuart and Shields 2017.

farmers were sent paperwork from the Murray–Darling Basin Authority to try to explain it. The Murray–Darling Basin Authority consultation period began on 28 November 2011 and ran until 16 April 2012.[93] Some of the largest public meetings were attended by a few thousand people. Kirkup remembered that government officials tried to explain the plan, but it was never clearly understood, and the meetings often resulted in even greater confusion. According to Kirkup, the government had never approached them in this dictatorial way. Until then, the farmers had believed that their water rights were secure, but the government's approach indicated a complete change in policy. The government's approach generated tremendous confusion and distrust and damaged the relationship between farmers and government officials. Kirkup recalled that everyone was terrified that the government would acquire all rights to the water.[94] The timing and approach of the meetings added to the feeling that the government was taking punitive action against the farmers. The meetings led to large protests and the burning of the Murray–Darling Basin Authority plan in a bonfire. There were, in fact, no compulsory acquisitions included in the plan, but at the time, there was a great deal of uncertainty about what acquisitions would look like.

The tensions between the farm groups and the Murray–Darling Basin Authority are just one of the social barriers identified in my research. Social tensions also exist between farm groups and urban-based environmentalists. Russell James of the Murray–Darling Basin Authority acknowledged that some urban environmentalists have a lot of influence but do not necessarily understand rural environmental issues. He explained, "the more active environmental groups tend to be urban based. For this reason, there is a risk of chardonnay-sipping types … or the city-based people who actually don't understand the river system." But his criticisms were mainly directed at farmers who he thinks are only interested in protecting their farm businesses. "We get a lot of backlash from the farming community in terms of the influence of environmental groups. Farmer groups have asked why the environmentalists are getting into hearings when they aren't the

---

93   Murray–Darling Basin Authority 2019a.
94   B. Kirkup, personal communication, 2016.

ones with millions of dollars invested in farm businesses." He then commended what he sees as a "reasonable" fraction of farmers who support the plan: "there's also a reasonable body of farmers that are actually what you might term 'green', or environmentally aware and so they are very strong activists for a balanced approach to things and do support the Basin Plan".[95]

Such comments have a dichotomising effect among farmers and government officials, among farmers and urban environmentalists, *and* within the farming community itself as it encourages social fragmentation between supporters of the plan and those who resist. Social dynamics lend legitimacy to some actors while undermining the legitimacy of other actors. As was the case of Griffith, social tensions embedded in the larger historical narrative had a significant effect on communications.

As Australian rural communities contend with ever-greater environmental challenges, addressing the social discord that makes these types of negotiations all the more challenging will be essential. An alternative community-based model that seeks to address these concerns is explored in detail in Chapter 5. As the above examples illustrate, it is imperative to pay attention to the underlying assumptions embedded in social relations among parties, as these have a real effect on the democratic process. Otherwise, important democratic negotiations can appear as window-dressing instead of truly meaningful engagement.

Finally, there were time barriers that affected the democratic process. Democratic pragmatism focuses on meeting certain consultative markers, like meeting with all the relevant actors. But the more authentic forms of participation can take time. Pressures to show progressive change within tight political timeframes can limit the capacity for comprehensive political participation to occur, resulting in a process that resembles an attempt to tick the boxes required by participation rather than to engage in deeper processes. Further, the farmers were under pressure to recover their businesses after ten years of drought. Such pressures meant that they, too, were constrained by time.

---

95   R. James, personal communication, 2016.

Time constraints led to decisions to release water flows without carefully examining the potential consequences for farmers. In one case, the hurried process wreaked havoc on the farm of Louise Burge. She pointed to the government's expedited decision to release environmental water flows as the cause of flooding and significant damage to her crops. In 2010–11, when years of drought finally broke, the river system was naturally relatively high due to significant unregulated flow from creek systems in the mountains. The river system's water levels, which would typically have dropped down in November and December, remained unnaturally high. The Murray–Darling Basin Authority and the state governments decided to put more environmental water down the system so water levels in the creeks did not drop. When Burge had to harvest the wheat on the other side of the creek, she could not get the header across. When they finally succeeded, they could strip just a few header loads before five days of rain caused them to lose all the remaining wheat. Their overall losses exceeded $350,000. At the time, they were never told that water would be put down the system, yet they received no compensation for the resulting losses.[96]

Burge ended up with one of the most significant personal losses in the whole of New South Wales because of one decision to artificially raise the river at Christmas time in the middle of harvest. As a result, she became one of the most vocal critics of the Murray–Darling Basin Authority. She told me that eventually, all the relevant ministers knew of their situation, and many even acknowledged a mistake, but no one would pay any compensation.[97] At the time of our interview, the Burges were still fighting to get the flow levels back to realistic levels and get a bridge or a crossing built so they could access half their farm when the creek's water levels were raised. In 2016, shortly after our interview, much of the Burge farm was submerged by flooding. The Murray–Darling Basin Authority had again released excess water from the dams without consulting the farm community.

The expedited process of establishing on-farm efficiency programs, with little time for consultation, also significantly affected farm

---

96   L. Burge, personal communication, 2016.
97   L. Burge, personal communication, 2016.

communities. From my research, it became clear that the water recovery goals of the Murray–Darling Basin Authority were simply not possible as articulated in the plan, particularly within their politically motivated timeframes. While the government could buy up significant entitlements, the plan's overall goals could not be achieved without projects that could deliver equivalent environmental outcomes. Further, the farming community was behind the push for efficiency projects. The government's timeframes for these projects were perceived to be based on politics and not reality. When it came time to develop the projects, state governments did not have enough time to do so – this was especially true for New South Wales, because it signed on to the basin plan later than the other states. There was simply no way to prepare multimillion-dollar projects with due diligence within unrealistic political timeframes.

There is often a political imperative to show that governments are taking expedient steps to address a problem. At the same time, there are consequences for rushing consultative processes. The cases illustrated above demonstrate that it is not enough to have democratic processes in place; there must be time to ensure those processes work to achieve their goals without profoundly detrimental consequences for those most affected.

Democratic pragmatism was associated with a narrow form of public engagement in the Murray–Darling Basin. As I have argued, this engagement did not substantially affect political decision-making. Instead, it systemically marginalised farmers. This marginalisation can be observed as a productive effect of a narrow reading of public engagement. Democratic pragmatism was limited in its consequences because it represented a shallow engagement effort. However, the application of democratic pragmatism revealed internal disciplines associated with the discourse coming to the fore. While efforts were made to engage in democratic consultations, for example, several barriers to participation were either overlooked or dismissed. The discourse of democratic pragmatism in the Murray–Darling Basin has internal disciplines that hinder the capacity of policymakers to engage in a genuinely democratic process. Ultimately, democratic pragmatism, with its adherence to the status quo, revealed how administrative

Table 3.3 Democratic pragmatism: transcripts and policy, legislation and actions in the Murray–Darling Basin

Transcripts, metaphors and rhetorical devices	Policy, legislation and actions taken
stakeholder engagement strategy	River Murray Commission (1917)
ticking the consultative boxes	Murray–Darling Basin Ministerial Council (1985)
committee-type approach	
leadership through funding	Murray–Darling Basin Agreement (1992)
public consultation process	
consultation	Basin Community Committee
outreach	Basin Officials Committee
bureaucracy	Murray–Darling Basin Authority–led town hall meetings

rationalism provided the overarching framework for action in the Murray–Darling Basin, as summarised in Table 3.3.

Conclusion

Viewing Murray–Darling Basin management through administrative rationalism helps explain the government's largely top-down approach to environmental water management. The discourse of administrative rationalism emerged to deal with the earliest water management challenges in the basin; these were challenges related primarily to salinity from the extensive irrigation systems. Management priorities shifted over time, but administrative rationalism's underlying assumptions and practices remained largely the same. As we have seen, administrative rationalism has considerable limitations, manifest in the disciplinary and productive effects of the discourse itself. These limitations are evidenced by an inability to seek out and incorporate locally based knowledge in decision-making and in a language of government that privileges the knowledge of "experts" over that of local populations.

Over time and corresponding with changes in management regimes associated with neoliberalism more broadly, some of the assumptions of Dryzek's economic rationalism are apparent in the problem definition and policy instruments enacted in the Murray–Darling Basin. But the evidence shows that there was no wholesale adoption of market-based tools by governments and other Murray–Darling Basin actors, as Dryzek suggests has occurred in other contexts.[98] Instead, the adoption of market-based instruments occurred mainly within the administrative rationalist frame. The government still led economic reforms, and regulations heavily influenced water trading. The Commonwealth government became the largest buyer and seller of water in the Murray–Darling Basin in the wake of management reforms, further centralising government control over water. As a result, some of administrative rationalism's underlying limitations or weaknesses were compounded and exacerbated by the addition of these market-based tools. These tools also came with their own problematic productive and disciplinary effects. For instance, a market-based system places a high value on *all* water. This high valuation meant that previously unused water suddenly entered the market as a tradeable "product". Water that had sat on farms and provided bird, fish and other wildlife habitat was now redirected towards more profitable ends.

The adoption of economic rationalist terminology by governments did lead to a new discursive response from farmers. Terms like "entitlement", "high value" and "risk" all have contested meanings. The government uses these phrases to enforce the dominant paradigm, and the second section of this chapter explained the productive effects of these terms. At the same time, farmers have adopted some of these terms to challenge how the state interprets economic rationalism. The result was the emergence of a discourse of resistance couched in the language of economic efficiencies. Using this language was sometimes effective in drawing attention to farmers' issues.

The analysis of democratic pragmatism showed that Murray–Darling Basin management attempted to create space for democratic engagement in the form of consultations, but these were minimal and largely inconsequential. What Dryzek referred to broadly as the problem-solving discourse of democratic pragmatism appears to have informed the inclusion of tools and practices that engage citizens. The evidence showed that consultations did not significantly affect

---

98   Dryzek 2013.

decision-making and were a far cry from more consequential and deliberative forms of democratic engagement that farmers and other critics of Murray–Darling Basin management believed should be in place to shape decisions. In other words, while the discourse of democratic pragmatism was employed by the governments (and sometimes by farmers), consultation tools were often communication tools of government rather than tools for democratic engagement. The adoption of these tools resulted in increased suspicion, an increased disconnect between farmers and government, and ultimately a farmers' discourse of resistance that demands more inclusion and accountability to the people who live in the Murray–Darling Basin, as will be discussed in greater detail in Chapter 5.

We can see a nod to economic rationalism as demonstrated by the prioritisation of individualism over community outcomes and a nod to democratic pragmatism as displayed through the limited mechanisms for democratic influence by key stakeholders. This constellation of discursive factors was critical in creating a misalignment between what the Murray–Darling Basin management plan under the Commonwealth *Water Act* of 2007 set out to do and the environmental results for the basin and for the human and non-human communities that inhabit it. These discourses (and their disciplinary and productive effects) and practices (such as consultations, town hall meetings, stakeholder input requests) then met with the worst drought in Australia's modern history to inform the creation of a new management plan for the Murray–Darling Basin. The Murray–Darling Basin Plan, which entered into law in 2012, was rooted in the discourse of administrative rationalism (with the addition of market-based instruments) but also included very particular ontological assumptions about the human–nature relationship – specifically the construction of "environmental water" and related concepts. This story is the focus of Chapter 4, which considers the productive effects of green environmentalism in the Murray–Darling Basin.

Ultimately, this chapter shows that while different environmental concerns arose over time, they were always dealt with through assumptions associated with administrative rationalism. Elements of economic rationalism, such as prioritisation of individualism over community outcomes, and of democratic pragmatism, such as limited mechanisms for democratic influence by key stakeholders, were present but administrative rationalism represented the overarching discourse. This was critical in creating a misalignment between what the Murray–Darling Basin management plan set out to do and the environmental results for the basin, including impacts on the human and non-human communities that inhabit it. The story of how the affected farming communities responded to this situation is the focus of Chapter 5. Before we get there, Chapter 4 discusses one final discourse, green environmentalism, to help explain what occurred in the Murray–Darling Basin.

Plate 1 Algae blooms on shore of the Coorong. All photos courtesy of the author.

Plate 2 Algae blooms in the Coorong.

Plate 3 Wildflowers in the Coorong.

Plate 4 Kangaroo tracks – salt lake in the Coorong.

Plate 5 Salt on the edge of the Coorong.

Plate 6 Salt lake in the Coorong.

Plate 7 Farm near Griffith, NSW.

Plate 8 Lake on a farm near Griffith, NSW.

Plate 9 Rice fields near Griffith, NSW.

Plate 10 Hay fields near Griffith, NSW.

Plate 11 Flooding downstream of Hume Dam.

Plate 12 Hume Dam.

Plate 13 Goolwa Barrages at the lower Murray.

Plate 14 Flooded farm near Wagga Wagga.

4

Green environmentalism

The environmental problems facing the Murray–Darling Basin are numerous. Most of the wetlands in the basin are threatened by increasingly long periods of drought between ecologically imperative floods. There are hundreds of forests in the floodplains dying from lack of water. Waterbirds that require certain water levels for breeding are threatened with extinction. Native fish populations are rapidly declining, toxic blue-green algae blooms are ever more common and salinity is becoming increasingly problematic.[1] As set out in the basin plan, the Commonwealth government's response has been to redirect water from productive environments, like farms, to environmental purposes. As Chapter 3 showed, a top-down, technocratic and science-based approach can be explained by the historically prevalent discourse of administrative rationalism. Another crucial piece of this story is about how "nature" has come to be understood and managed in Australia through a discourse that I call "green environmentalism".

This chapter shows how green environmentalism initially emerged as a challenge to administrative rationalism (and its associated projects like large dams) through a specific understanding of "nature" as separate and distinct from humans and their productive uses of land. This biocentric understanding of nature, which excludes people, is a

---

1   Pittock and Connell 2010, 564.

departure from the other discourses. As the green discourse evolved, it became embedded within the established political order through the Australian Greens. But green environmentalism came to be practised in a way that reflected the overarching dominance of administrative rationalism. The implementation of green environmental practices also reflected (to varying degrees) the values extolled by economic rationalism and democratic pragmatism. In sum, even though the aims of green environmentalism were originally distinct from the other discourses, the discourse has come to reflect many of the assumptions embedded in the other dominant discourses. For example, green environmentalism came to embody aspects of administrative rationalism, including its emphases on "expert" planning, bureaucratic design and top-down solutions. The problem definitions of green environmentalism reflect these ideas, as highlighted in common transcripts such as "environmental protection" and "environmental water". What remains distinctive about green environmentalism is the core assumption that economic growth is inherently exploitative of natural environments. Therefore, the way to protect nature is to create separate spaces away from human interference. In this way, it is important to explore green environmentalism as a distinct discourse with unique productive and disciplinary effects.

Today, the overall resource management approach in the basin, encapsulated in the transcript "environmental water", separates ecological health from productive land uses such as farming. Consequently, problem definitions centre around the notion that the environment can only flourish if water is diverted from agriculture to nature. This chapter illustrates how environmental water and its associated management practices have certain productive effects. For example, the idea of environmental water constructs farms as landscapes that are only for raising crops and livestock and not as cultivated habitats that also bring benefits to wild species. "Nature" is seen as something that does not require active management. Further, environmental water has the disciplinary effect of defining policy solutions narrowly, namely by assuming that, despite high natural variability, certain levels of water are "normal", that "nature" needs definite amounts for different purposes and that this water must come from farming to achieve environmental goals. This discourse also has significant disciplinary effects, such as conflating many ecological concerns with inadequate water volumes in rivers and dismissing efforts by farmers to increase the efficiency of water use in agriculture. Green environmentalism informs environmental reform in the basin, and "environmental water" encompasses how government problematises water management. As a problem definition, the conceptualisation of environmental water informs the kinds of policy prescriptions considered appropriate and acceptable to governments. In this way, green environmentalism, as exemplified by the transcript of "environmental water", has significant productive and disciplinary effects that are not explained by the other discourses I have examined so far.

Green environmentalism in the Murray–Darling Basin

The environmental movement in Australia is best characterised by an ontological understanding that humans are separate from nature, a view that some Indigenous scholars are now challenging.[2] This biocentric discourse fails to recognise that human beings have always had significant impacts on their environment.[3] As such, the discourse has the disciplinary effect of downplaying the positive impacts that productive environments can have on the land. In Australia, the Commonwealth government's primary response to the Millennium Drought (1997–2009), as set out in the Murray–Darling Basin Plan, was to redirect water from productive environments like farms and set aside water for environmental purposes. This response has been influenced by the nature preservationist movement, which emerged in the 1970s.[4] Australian preservationists drew inspiration from the American wilderness and biodiversity conservation narratives,[5] which expressed ecocentric values in an anti-materialist stance towards the industrialisation of nature. This movement now defines the bounds

---

2   Pascoe 2018.
3   Bookchin 1994.
4   Christoff 2016.
5   Classens 2017.

of the modern Australian environment movement.[6] However, the movement's conceptualisation of non-human nature is increasingly problematic.

The environment movement in Australia emerged in the 1960s in response to widespread environmental degradation, pollution and pressures on the natural world. From the 1960s through the 1970s, in the wake of Rachel Carson's famous book *Silent Spring*, campaigns were initiated to protect natural environments against development threats. The first green party in the world was the United Tasmania Group, which was formed by members of the Lake Pedder Action Group. The Lake Pedder Group had sought to challenge the removal of national park status and protest the flooding of the south-western Tasmanian lake as part of a major hydroelectric project in the late 1970s.[7] The campaign against a dam that would have flooded the valley of Tasmania's Franklin River mobilised tens of thousands of Australians in the 1980s and helped solidify the base of Australia's environmental movement. Bob Brown, who became the leader of the Australian Greens in 2010, was a founding member of the Tasmanian Wilderness Society and its director in 1978. By 1982, 1,500 Franklin Dam protestors had been arrested, including Brown.[8] Shortly after, in 1984, the Tasmanian Greens was formally registered as a political party.

Environmentalists also focused much of their energy on the Great Barrier Reef, hoping to protect it from mining and oil drilling. Community-driven support for the protection of the reef garnered national attention. During the 1980s, as the environmental movement grew to enjoy wide support, the environment became a major political issue. After extended campaigning, some of Australia's most significant wilderness areas were granted protection, and the Commonwealth government asserted its constitutional power to protect sites of World Heritage.[9] Bob Brown co-wrote *The Greens* in the mid-1990s. The book sought to define the values of the party. Brown advanced a largely ecocentric philosophy. He wrote: "I am not a conventionally religious

---

6   Christoff 2016.
7   Kerr 2013.
8   Kerr 2013.
9   Australian Environmental Grantmakers Network 2014.

man, but in the wilderness, I have come closest to finding myself and knowing the universe and accepting God – by which I mean accepting all that I don't know."[10] The origins of the Greens as a conservation-based movement led to their predominantly ecocentric philosophy. Hillier wrote: "The result was the development of an eco-centric philosophy that was, initially, central to green politics. An unstable amalgam of the romantic (which celebrates diversity, emotion, and the encounter) and the scientific (which claims universality, law and rationality)."[11] The preservation of the natural world was central to the political ideology of the movement and the early development of green politics:

> In Australia, the practice of wilderness preservation, and the theoretical endeavour of environmentalists to establish a body of thought justifying this practice, was central to the early development of Green politics. Many environmental activists were originally moved by the emotional and aesthetic impact of the "natural world"; determined to save it from human interference.[12]

The green movement was a strong reaction to the economic environment of the 1970s and 1980s. Unfettered economic expansion at the cost of the environment had devastating consequences. Efforts to protect areas like the Great Barrier Reef, for example, were critical. Nonetheless, the green movement grew out of Tasmania, an area of Australia with a very different history of development from the mainland. The history of intense irrigation and European-style cultivation that had dominated New South Wales for more than a hundred years was much different from that in Tasmania, where intense dam development did not begin until the 1970s. Nonetheless, the Greens managed to gain significant support on the mainland, mainly due to support from urban constituents.

---

10   Brown and Singer 1996.
11   Hillier 2010, 2.
12   Hillier 2010, 1–2.

The federal election of 2001 represented one of the first major breakthroughs of the Greens. Their rise (to 5 per cent of the popular vote) reflected dissatisfaction with the Labor Party and the Liberals. Both parties generally showed little political will to counter some of the more detrimental neoliberal reforms of the period. The Greens were able to garner support from middle-class voters by campaigning on social issues, which represented a shift away from the party's focus on environmental issues.[13] While in the past, Brown, for example, had argued that stopping population growth (including by way of immigration) would be essential for environmental protection, the refugee crisis compelled the Greens to take a more socially liberal approach to immigration and resettlement planning.[14] The more progressive and vocal stance on immigration garnered the support of a broader voting base as the party was no longer seen as a single-issue party.

Changes in Australia's electoral system secured the power of the Greens. Preferential voting under proportional representation gave the Greens the capacity to influence and even determine the outcome of the electoral contest between Labor and the Coalition (the Liberal–National coalition is an alliance of centre-right political parties in Australia and one of the two major political forces). According to Manne, the rise of the Greens reflected "an expression of shifts in social consciousness or unresolved tensions in the political culture".[15] The national Greens party was formed in 1992. The party's support base came mainly from inner Sydney and Brisbane, where questions of social equity, urban development and democracy were more central than wilderness preservation.[16]

In the late 1990s, the Australian Conservation Foundation, Australia's foremost non-profit national environmental organisation, focused the spotlight on the Murray–Darling Basin and gained the support of the Greens. Under the leadership of Peter Garrett, the ACF tried to change public perceptions about the sustainability of

---

13   Hillier 2010, 13.
14   Hillier 2010, 14.
15   Manne 2010.
16   Hillier 2010, 7.

agriculture in the Murray–Darling Basin.[17] In 2000 the Australian Conservation Foundation published a report called *National Investment in Rural Landscapes*. Funded by the Land and Water Rural Research Development Corporation, the report asserted that "the annual cost of degradation in rural landscapes is at least $2 billion annually, and this figure is rising. With no action, this could balloon to over $6 billion annually by 2020." The report advised that a capital investment of $60 billion, with an ongoing maintenance program of $500 million, was required over ten years. These amounts represent a total investment of around $6.5 billion per year. The foundation's arguments also appeared in a 2001 report called *Repairing the Country* by the Allens Consulting Group. The World Wide Fund for Nature, along with the Wentworth Group of Concerned Scientists, soon joined the campaign for the Murray–Darling Basin.[18]

Under pressure from the green movement and calls for reform in the wake of the Millennium Drought, the government of Australia effectively took control over water rights in Australia by enacting the *Water Act* and guaranteeing 2,750 gigalitres for the environment.[19] Citing its responsibility under the Ramsar Convention, its capacity to access the best available science, to manage water markets across state boundaries and to conduct appropriate socio-economic studies, the Commonwealth government took greater control over water resources through the Intergovernmental Agreement on a National Water Initiative in 2004. The states, through the Council of Australian Governments, ceded control over water to the Commonwealth government. According to the national government, state governments acknowledged that water is an issue of national significance. The National Water Initiative states that the 1994 Council of Australian Governments water-reform framework and subsequent initiatives indicate that Australia's water resources would be better managed under the Commonwealth government. The Commonwealth government focused on areas they deemed largely unaffected by

---

17 Nahan 2003.
18 Murray–Darling a threatened river 2007.
19 Australian Government 2024.

development.[20] Since the drought, the Commonwealth has focused on protecting wetlands by taking water out of agricultural production.

The Murray–Darling Basin received support for this approach to water reform from a wide range of actors, including scientific bodies, conservation groups and academics. The Commonwealth Scientific and Industrial Research Organisation (CSIRO) allied with the Murray–Darling Basin Authority.[21] The green movement, with direction from the Australian Conservation Foundation and the World Wide Fund for Nature (WWF), provided the drive towards Commonwealth government–led water reform in the country. In addition, several scholars of the Murray–Darling Basin supported this approach, arguing that the Commonwealth government was best positioned to manage the crisis.[22] As evidenced by the involvement of a wide range of environmental actors, the Commonwealth government gained widespread support for its approach to Murray–Darling Basin management.

Green environmentalism leads to a particular way of understanding and defining problems. For example, in policy terms, the Commonwealth government constructed the idea of environmental water to differentiate between water intended for human use and water intended for environmental use. According to the South Australian government, for instance, water for the environment is allocated to meet the ecological needs of plant and animal communities to survive and reproduce or "water allocated purely to the environment and not extraction".[23] The water is used explicitly for nature, independent of

---

20 It is interesting to note that ecological character, as described by the Ramsar Convention, is not exclusive of human managed systems. At any point in time, regardless of past human interventions, sites can be listed as protected sites. While the green movement in Australia emphasises "nature" as separate from people, Ramsar does not make these kinds of distinctions and has protected several wetlands that have been heavily affected by human activity, including wetlands that serve as rice paddies in Vietnam.

21 Murray–Darling Basin Authority 2021d.

22 Grafton and Horne 2014; Pollino, Hart et al. 2021; Ross, Buchy and Proctor 2002.

23 Government of South Australia Department of Environment and Water 2016.

people. The Murray–Darling Basin Authority calls environmental water "water for the environment":

> [which is] used to improve the health of our rivers, wetlands and floodplains. Water is allocated to federal and state environmental water holders across the Basin, who make decisions about when, where and how much water is released for the environment, with measurable environmental outcomes in mind.[24]

These definitions separate environmental water from water used for agricultural or productive purposes and seek to define water for environmental purposes as distinct from water used on farms.

Calls to save the Murray–Darling Basin from the impacts of agriculture are now widespread, and academic literature on water management in the Murray–Darling Basin tends to assume that farmers are unconcerned about environmental impacts. For instance, Mallawaarachchi and colleagues wrote: "Recurring droughts and resultant scarcity of water has made negotiations further complicated and controversial, broadening the gulf between environmentalists seeking public good outcomes and irrigators seeking private profit."[25] Some researchers charged farmers with only being concerned with profits, while others argued that farmers were primarily responsible for the environmental problems in the basin.[26] But questions of whether humans are part of nature and whether human actions and their consequences are "natural" call into question the preservationist ethic central to green environmentalism. These questions are increasingly important for modern environmentalism to revisit as technological innovations and the effects of human actions commodify and reconfigure the non-human world at every level.[27]

---

24 Murray–Darling Basin Authority 2019d.
25 Mallawaarachchi, Auricht et al. 2020, 345.
26 Crase, O'Keefe and Dollery 2014 and Ross, Buchy and Proctor 2002 (on concern for profits); for responsibility for problems, see Grafton and Horne 2014.
27 Christoff 2016.

Although the environmental movement in Australia appears to have yet to wrestle with these issues, nature has long been "denaturalised" in academic circles. Since the 1990s, the term "nature" has been problematised by various thinkers.[28] These scholars have argued that nature is inseparable from human communities. While the pristine natural beauty of the countryside is valued, productive and natural spaces are no longer distinct. As argued by Vanclay and Lawrence in *The Environmental Imperative*,[29] for farmers, the natural world is not external to the social world. The social and ecological problems in the basin are inseparable. From this perspective, productivity does not necessarily undermine the natural ecology of the land.

In its campaigns to save ecosystems and species from the impacts of destructive industrial practices, the green movement's staple discourse largely remains a preservationist one, despite the turn towards a denaturalised nature in academic writing. This movement still builds on the romantic tradition of valuing pristine nature – a nature essentially unchanged by humans.[30] Yet colonisation and Indigenous settlement transformed Australia's natural environment.[31] With climate change further refashioning ecosystems at an accelerating rate, it is hard to understand precisely what the environmental movement is now trying to protect. The movement's seemingly straightforward relationship with nature is growing increasingly uncertain, and its identity as nature's defender is increasingly unstable.[32]

In sum, green environmentalism has become the overarching paradigm that informs environmental reform in the basin, and environmental water has become the dominant policy response to water management problems. But, as this work argues, the paradigm is problematic as there is increasing uncertainty regarding what exactly

---

28　See for example Cronon 1996; McKibben 1989; Oelschlager 1991; Soper 1995; Vogel 2015.
29　Monk 1997.
30　Christoff 2016.
31　Pascoe 2018.
32　Christoff 2016.

needs protection. While recognising the valuable contribution the green movement has made towards Murray–Darling Basin management, the discourse has effectively undermined the critical role farmers can play in environmental management. My analysis provides a clear case study of the impacts of an environmental discourse that separates nature from human communities. Murray–Darling Basin Authority policy separates environmental water from water used for productive purposes on farms. This approach is rooted in the green movement in Australia and legally backed by Australia's signatory status in the Ramsar Convention on Wetlands of International Importance Especially as Waterfowl Habitat. Several examples in this chapter illustrate how the green perspective affects problem definitions and policy prescriptions and is often at odds with the perspective of farmers who find it impossible to understand the separation of human and "natural" environments. This distinction is explored in this chapter and the next. The latter sheds light on how farmers see the human and ecological aspects of communities as inseparable. But first, I turn to the Ramsar Convention. This international treaty provided much of the legal justification for the Commonwealth government's protectionist approach. While the focus on wetlands protection was an important and welcome development, it also reinforced biocentric arguments that downplayed the positive role that farming communities could play in managing water resources in the basin.

The role of Ramsar in establishing the basin plan

The Greens shaped the modern Australian environmental movement as a political force. As a result, the party came to have a significant influence on public policy by the early 2000s. The Greens' push for water reform contributed to the development of the *Water Act* and the basin plan. Australia's signatory status as part of the Ramsar Convention (the international treaty for wetlands protection) served as legal justification for the decision to offer wetland protection. The Ramsar Convention is a multilateral environmental treaty signed in Ramsar, Iran, in 1971, coming into force in 1975. Citing its obligations under the convention and its access to the best available science, the

Commonwealth government gained more control over water resources through the 2004 National Water Initiative. The *Water Act* and the basin plan derive their constitutional validity from these international agreements. If the basin plan does not meet international obligations, it could be considered invalid.[33] The Ramsar Convention provided a critical enforcement mechanism for the basin plan and justified Commonwealth intervention in what is constitutionally a state-led policy jurisdiction. This change represented a reframing of the water management problem as a key priority and responsibility of the Commonwealth government.

According to the Ramsar Convention, a wide range of habitats can be classified as wetlands, both "natural" and those altered by humans. Wetlands include swamps, marshes, billabongs, lakes, salt marshes, mudflats, mangroves, coral reefs, fens, peat bogs and bodies of water. Waters within these environments can be static or flowing, brackish or saline, and can also be inland rivers and coastal or marine water to a depth of six metres at low tide. When a Ramsar site is established, countries form management frameworks to ensure its "wise use", broadly defined as maintaining the ecological character of the wetland. The "ecological character" is "the combination of the ecosystem components, processes, benefits and services that characterise the wetland at a given point in time". That given point in time is the point at which the site is listed, regardless of past conditions.[34] It is interesting to note that "ecological character", described by the Ramsar Convention, is not exclusive to human-managed systems. Despite past human interventions, sites can be listed as protected sites at any point in time. Thus, while the green movement in Australia emphasises "nature" as separate from people, Ramsar does not make these kinds of distinctions and has protected several wetlands that have been heavily affected by human activity (particularly wetlands that also serve as rice paddies in Vietnam). Nonetheless, the Australian government focused on areas it deemed largely unaffected by development.

As discussed in Chapter 1, the government effectively took control over water rights by enacting the *Water Act* and guaranteeing 2,750

---

33   La Nauze and Carmody 2012.
34   Australian Government Department of the Environment and Energy 2019.

gigalitres for the environment.[35] The state governments ceded control over water to the Commonwealth government but was able to negotiate – to some degree – what the reform process would look like. While this change was presented as politically neutral, some farmers I spoke with argued that giving greater powers to the Commonwealth government undermined the Australian Constitution, which accords water rights to state governments.[36]

The 2004 Intergovernmental Agreement on a National Water Initiative claimed that the 1994 Council of Australian Governments (COAG) agreed that managing Australia's water resources is a national issue. The agreement says that, because of this recognition, states and territories have made considerable progress towards more efficient and sustainable water management. States and territories embarked on a significant program of reforms to their water management regimes. Regulatory measures, such as setting extraction limits in water management plans and specifying the conditions for the use of water in water-use licences, were undertaken. These jurisdictions would also continue to examine the feasibility of using market-based mechanisms such as pricing to account for positive and negative environmental consequences associated with water use. Further, they implemented pricing that includes environmental costs (referred to as externalities in the agreement), where it was "found to be feasible".[37] Directing the course of reforms, the Commonwealth government situated itself as best able to make decisions based on science and socio-economic studies. Using the Ramsar Convention as further justification, the Commonwealth government brought water management largely under its control.

The question of what constituted an environment and the details of the government's environmental goals were essential to farmers who understood the plan's potential impact on their communities. The modern history of the Murray–Darling Basin is characterised by broad and far-reaching changes to the environment for the purposes of development. Therefore, it was unclear to farmers what the

---

35   Murray–Darling Basin Authority 2014.
36   Commonwealth of Australia Constitution Act 2019.
37   COAG 2004.

government's end goal was in restoring natural systems. The Murray–Darling Basin is a highly modified system, with much of the land converted to productive irrigated agriculture over the last hundred years. While there are many marshlands, most were established after the dams were built and the river system was modified. Further, the lower lakes were converted from saltwater to freshwater estuaries with the construction of weirs. While many people conceive the environment as a natural space free from human interventions, the reality in most of the basin is quite different. Farmers challenged the government's view that some sites – and not others – deserved environmental protection under the Ramsar Convention.

One of the most significant critiques made by farmers was that the plan did not consider the environmental value of farms. They saw their farms as part of Australia's natural landscapes and themselves as caretakers within those environments. They resented that so much water was being taken out of the system for wetlands and environmentally protected sites when their farms demonstrated many of the same attributes as these protected sites. Farmer Debbie Buller remarked: "this whole idea that agriculture and environmentalism are on opposite sides of the spectrum is insane. Aren't we part of the environment?"[38] Farmers said they considered themselves environmental stewards, caring for the abundance of wildlife on their farms, including frogs, birds and other species. Further, it was evident that many of the natural environments the Commonwealth government wanted to provide water to were contested in terms of their status as natural environments. The Ramsar Convention, as justification for these decisions, was often called into question. In what follows, I explain how farmers are constructed in green environmental discourse. I then show how farmers challenge this dominant frame from within green environmental discourse while sometimes challenging it outside the discourse.

---

38   D. Buller, personal communication, 2016.

Constructing "farmers" in green environmental discourse

Green environmental discourse has generally characterised farmers in a negative light. Such characterisations have significant productive and disciplinary effects on problem definitions and policy solutions. Farmers, consequently, seek to resist this discourse from both within the discourse itself and outside it. While there are certainly farms with a large ecological footprint, there is a tendency to lump all farms together in terms of thinking about environmental effects, a view that farmers seek to challenge. Russell James of the Murray–Darling Basin Authority acknowledged that while the farm community is not a homogenous group, he also believed that some farmers are adamantly opposed to the plan. He understood their opposition to the plan as a disregard for environmental interests. He told me: "There's some very big players in the industry, who are just pushing for more and more water to be available and make more money. They couldn't give a stuff about the environment, and they would like the government to pay for everything that they do." At the other end of the spectrum, James told me, there are those he called "responsible farmers, who understand we live in a variable climate, understand we have droughts, understand those resources need to be looked after, that there's a limit to how much you can take out of it". He congratulated some farmers for taking the initiative to set up environmental water groups themselves to look at the system holistically.[39] While James recognised that farmers are not a homogenous group, and that some act environmentally responsibly, he expressed a widespread opinion that farmers do not care about the envrionment. He also conflates opposition to the plan as opposition to environmental initiatives generally.

For farmers, the perspective that they do not care about the environment reflects an urban disconnect with rural life. Farmer Allen Clark believes the divergence in views can be attributed to a lack of understanding on the part of city people about the challenges of farmers and their way of life.[40] In previous eras, most people knew someone living on a farm, but today most people who live in cities do

---

39   R. James, personal communication, 2016.
40   A. Clark, personal communication, 2016.

not visit farms. Further, city dwellers tend to view farms as exclusively productive spaces. The rift in the views of city dwellers and farmers is readily apparent in the remarks of farmers who report feeling verbally attacked and are prone to do their share of attacking too. Farmer Debbie Buller joked: "in the city, you have Gladys Stringbag, who votes for the Greens and says those bastard farmers are using all the water, and there is nothing for the environment". Buller explained that such an attitude implies there is no environment where the farmers are and that there is a clear delineation of environment and agriculture. Buller attributed the lack of understanding of rural farm communities directly to the rise of the green movement. When she and her husband Stuart started farming, representatives from the CSIRO used to come to the farm and, wearing their gumboots, work with them in the paddock. She told me that does not occur any more.[41] Buller believed that the government's change in approach is related to the view of the green movement that farms are not environmental spaces.

Those in farm communities tend to believe that people in cities, particularly government representatives, do not understand the relationship that farmers have with the land. During the visit of a government employee to Buller's farm, Buller's husband picked up a handful of soil and smelt it. The government employee asked Buller what he was doing. She explained that he can assess the quality of the soil from the smell: "there is a particular microbe in the soil that gives it a certain smell, and the farmers can check the biology of the soil by the smell".[42] Buller added that farmers care deeply and even "worship" the land.[43] When talking about the environment, it is evident that Dalton and Buller appeal to the romanticism prevalent within green discourse. The farmer tries to accentuate their deep connection to the land and reframe themselves as environmentalists. This framing is an example of farmers' resistance within the established discourse. The green perspective focuses on romantic notions about natural spaces, and the deep connection people can share with the land. Farmers appeal to this dominant green discourse to undermine the authority

---

41 D. Buller, personal communication, 2016.
42 D. Buller, personal communication, 2016.
43 D. Buller, personal communication, 2016.

of government officials by elevating their connections to the land. For example, farmer Helen Dalton remarked, "you have to get a bit of dirt under your nails and understand a bit about the land".[44]

Farmers like Louise Burge seek to redefine themselves as environmentalists by using the language common to green environmental discourse. Burge supports any kind of environmental policy that stops destruction: for example, protecting areas from urban sprawl or mining. These comments reflect the protectionist ethic that originally defined the green movement. She believed that regarding farms, another approach to the environment is warranted because there are elements of natural environments within agricultural landscapes. In this way, Burge challenged aspects of the green environmental discourse from within the discourse itself. She sought to reposition farms as part of the natural system and farmers as environmental stewards. Further, she wanted environmentalists to see the need for human interventions in natural systems, thus presenting a direct challenge to green environmental discourse. In the case of natural bushland settings, for example, she argued that there is a need for human interventions like controlled burning. While people think they must protect nature by "locking it up and leaving it alone, this can be a terrible mistake".[45]

The value of farming is lost in the space that exists between consumers and producers. Like in Karl Marx's critique of commodity fetishism, the social relationships between people are reduced to economic transactions, and both the consumer and the producer become abstract concepts, far removed from the mind of the other as part of their concrete reality. My research points to this ever-widening divide, which leads to many misconceptions about the "other". Contrary to common perceptions, my study suggests that farmers share many of the same environmental values as city dwellers. Still, they question the way environmental values and natural spaces are defined. Negative perceptions, whether of farmers or city dwellers, have a dichotomising effect on policy recommendations and outcomes in the Murray–Darling Basin. The green discourse has defined water as either

---

44   H. Dalton, personal communication, 2016.
45   L. Burge, personal communication, 2016.

an environmental or productive asset, not both. As such, the policy recommendations that stem from that approach tend to reinforce the already established rift between city dwellers and farmers, who both see themselves as environmentalists. In what follows, I demonstrate how green discourse has both productive and disciplinary effects on policy choices.

Farmers challenging the discourse

Farmers challenge the discourse in the Murray–Darling Basin in several ways. The Commonwealth government's position was that a significant increase in the volume of water available to the environment was essential to ward off environmental collapse caused by climate change. Farmers contested the science and argued that the Australian climate was subject to longer-term climate cycles than were considered in the plan's development. Both sides agreed that cyclical factors needed to be considered to determine what was "normal" and what actions were required. While the drought lasted ten years, farmers pointed to its eventual end and preceding years of heavy rain as evidence of cyclical patterns of drought and flooding. The government argued that drought would become the new norm because of climate change. While this appears to be a purely technical debate, it is significant for the future of the Murray–Darling Basin Authority, because defining "natural" climate systems provides the basis for understanding what future systems should look like. This is an example of how both parties work within the discourse of green environmentalism to advance their interests. The perceived goal is an ideal state of nature that precludes human interference.

Mike Makin of the Murray–Darling Basin Authority explained that while there is a cliché that Australia is the land of droughts and flooding rains, the variation of the river systems is among the highest in the world and so lends truth to the idea. But he believed the shift in climate over the last 40 years has altered the patterns and projections of what people are likely to experience.[46] Similarly, David Dreverman, also of

---

46 M. Makin, personal communication, 2016.

the Murray–Darling Basin Authority, believed that the extent of water scarcity has caused the Commonwealth government to reset all its water management arrangements, particularly around the dry sequences.[47] According to Makin and Dreverman, the script farmers use to describe Australia's climate, "the land of droughts and flooding rain", is an attempt to disregard efforts to improve the environment.[48] Farmers I spoke with rejected such claims. They questioned the plan's logic because the Commonwealth government's conception of the climatic system did not fit within farmers' understanding of Australia's environment. Neither group sought to undermine the basic assumptions embedded in the discourse: that nature has an ideal condition that we must strive to realise. Both groups also prioritise the importance of expert knowledge and scientific data in decision-making.

In line with administrative rationalism, both farmers and Commonwealth government officials sought to establish themselves as the experts on climate cycles. Hilary Johnson of the Murray–Darling Basin Authority told me that many farmers have said that they do not understand the problem the government is trying to solve, mainly because droughts are a natural part of the system in Australia. Johnson explained that the government had based its approach on science that shows a long-term decline in species and communities across the basin.[49] But the available scientific evidence relies on contested

---

47 D. Dreverman, personal communication, 2016.
48 This line comes from a famous poem by Dorothea Mackellar called *My Country*. The poem celebrates her love and connection with the Australian landscape.
49 This plan was prepared as a result of a public consultation and submissions process that showed a decline in species and in communities. The Murray–Darling Basin Authority prepared a draft plan, which was the basis of the public consultation process. A 20-week period was allowed to receive briefings on the draft plan, to attend round tables and public meetings and to prepare submissions in response to the plan. Almost 12,000 submissions were received. The Authority reviewed these submissions and made changes to the draft plan. A summary of the submissions, how the Authority addressed those submissions and the resulting changes to the proposed basin plan were published in the *Proposed Basin Plan consultation report – May 2012*. According to the process outlined in the *Water Act*, the Authority was also

assumptions. Johnson explained that the scientific evidence points to a long-term decline across water-dependent ecosystems. For Johnson, the problem was not managing drought but planning for the *potential* impacts of increasing drought.[50] Yet without conclusive evidence to determine the natural cycle, it is impossible to prove that present conditions are worsening. The current way of defining the problem, in line with green environmental discourse, tries to define what a "natural" system looks like. As evidenced in debates in the scientific community (as I will explain), there is currently no consensus on these natural systems.

Many farmers saw themselves as highly knowledgeable regarding climate patterns. The older irrigators told the younger irrigators that they had previously seen all these climate patterns, which perpetuated an atmosphere of doubt about climate change. Tree rings and other indicators helped scientists and irrigators determine cycles, but historical records were incomplete and only dated some two hundred years. No one predicted that, after the long dry period, they would have two major floods in two years.[51] There was no agreement in the farm community about the nature of climate change or if it was affecting the natural cycles of Australia's climate. While the economic implications of the plan certainly contributed to resistance, farmers were also so used to drought that they generally saw it as a normal condition that could be managed. Aside from the major variation in precipitation during the Millennium Drought or the "big dry", we see that there does not appear to be a distinct pattern regarding rainfall and climate in Australia. The unpredictability of climate, unclear scientific evidence and knowledge passed down generationally all fed into very different interpretations about the nature of climate cycles in Australia. These factors, combined, contributed to a view among farmers that the available evidence does not support the position or the initiatives undertaken by the government.

---

required to seek comments from members of the Murray–Darling Basin Ministerial Council. Ministers had six weeks to respond to the Authority on the draft plan (Murray–Darling Basin Authority 2019d).

50   H. Johnson, personal communication, 2016.

51   B. Kirkup, personal communication, 2016.

Farmer Louise Burge believes that the severity of the Millennium Drought can be compared to the Federation Drought of 1895 to 1903 and another major drought in the 1930s and 1940s, pointing to a cyclical pattern. Farmer Helen Dalton cited historical records that drought was unexceptional. For example, when James Cook first landed in Australia in 1770, the country was in the midst of a 20-year drought on the east coast. Helen Dalton's knowledge of droughts included one in the 1890s, the Federation Drought from 1902–03 and drought years in the 1940s. Further, there was also a drought in 1982.[52] Farmer Bernard Walsh believed that the Murray–Darling Basin is headed towards an overcommitment of the river system but took issue with the fact that the plans were drawn up in a "one-in-a-hundred-year" drought situation that no one had experienced before. Walsh guessed that the climate in Australia might be a one-in-a-hundred-year cycle because the last extended and severe drought was from about 1890. For Walsh, using the worst drought years as a benchmark for increasing water flows did not make sense.[53] These farmers believe, in several variations, that severe droughts are just part of the natural system in Australia.

The Commonwealth government's depiction of the drought problem made farmers even more resistant to their scientific conclusions. One prominent scientist, Tim Flannery, talked to farmers about climate change. According to Dalton, in one such meeting, Flannery told the irrigators' group that it would never rain again and the dams in the Snowy Mountains would never fill again. She recalled:

> Tim Flannery, he's got a book out called *The Weather Makers*, and he was the climate commissioner. He's an anthropologist. People prostitute themselves for money. He was paid to tell us about climate change, and we were dovetailing in this Millennium Drought, so we were believing them. It was a really grim time for everyone.[54]

---

52 H. Dalton, personal communication, 2016.
53 B. Walsh, personal communication, 2016.
54 H. Dalton, personal communication, 2016.

Remarks from scientists like Flannery caused great alarm within the community, with many farmers believing they would never farm again. The presentations of government-hired scientists were perceived in such a way as to cause significant discord. Some farmers responded by deciding to exit farming altogether, while others became entirely dismissive of scientists like Flannery because they painted such a fatalistic picture of the situation. The government may have intended to incite action and cooperation within the farm communities but instead created further divisions within farm communities and with the government. Such distrust of the government was further exacerbated when the end of the drought brought the heaviest rains in Australia's recorded history. During my interviews in 2016, the communities were reeling from extensive damage caused by flooding. The farms in the region had lost millions of dollars from flooding. The extensive rainfall, and the significant losses, further solidified distrust in the farm community about how much government officials understood climatic cycles in Australia.

It is difficult even to begin a reform process when there is no real agreement on the nature of the problem. Determining whether there is a change in the climate in Australia invariably affects whether people agree there is a crisis that requires action. If farmers were not convinced there is a change in the cyclical patterns of drought and rainfall, it would not be possible to convince them of the value of taking water out of the productive system for environmental purposes. This dilemma points to a pertinent question that is overlooked on both sides, with their focus on "natural" climate cycles: given the level of human intervention in the Murray–Darling Basin, how can we know what the natural cycle should look like?

Both sides of the debate are affected by the biocentrism prevalent in the green perspective. Both use scientific evidence to determine "natural patterns" and provide proof of what is valid. But my interviews with farmers and government officials suggest that decision-making about what is "natural" is often based on social or environmental values. There is a focus on determining normal weather patterns and cycles without recognising that those cycles have invariably changed due to the numerous human interventions that have occurred in the last one hundred years. Green environmentalism fails to acknowledge the impacts of human interventions and anthropogenic changes, and that reverting to an imagined ideal climatic system may be impossible. While the Commonwealth government recognises the effect humans might have had on climate cycles, it relies more heavily on evidence of climatic patterns from the past. A critical look at green environmental discourse reveals that it presents no straightforward way of reaching a consensus on ecological goals, despite scientific evidence of climate cycles or a lack of such evidence. While it is essential to address the impacts of drought, the answers may not lie in the past but in looking towards a future vision that embodies the goals and values of all those affected.

Like the debate around climatic patterns, contention grew around the government's focus on diversions, the most significant policy directive and monetary investment for addressing the crisis. While green discourse emphasised getting water to wetlands and protected areas, administrative rationalism led to controlling as much water as possible and tightly governing decisions around water releases. In 2012, desired environmental objectives were identified under the management plan for the basin's water.[55] To achieve these objectives, an additional long-term average of 2,750 gigalitres of water per year was set aside to restore and maintain the health of the river system.[56] A strong focus on water volumes was problematic, as it had the disciplinary effect of overlooking how water infrastructure contributed to water- and soil-quality problems in the basin in the form of salinity. Sufficient water flows are essential since 2 million tonnes of salt leach out of soil and rock every year and are released down the waterways of the Murray–Darling system. This salt must be flushed into the ocean, or it causes significant problems in the freshwater system. Without flushing flows, salinity and algae blooms increase, causing severe environmental issues.[57] But sending more water down the system did not adequately address salinity problems. Dams and weirs affect flows and reduce the effectiveness of flushing, while excess flows can cause blackwater events (which will be discussed in the next section).

---

55   Murray–Darling Basin Authority 2012.
56   Murray–Darling Basin Authority 2014.
57   Murray–Darling Basin Authority 2014.

Many farmers questioned the emphasis on water volumes and were concerned that other approaches to addressing salinity problems were being overlooked. Further, there are certain aspects of the system that, according to many farmers, make flushing impossible. While farmers play a vital role in managing water in the Murray–Darling Basin, it must also be understood that the system as it exists today, and all the problems associated with it, can be attributed to the fact that the system was modified to support the kind of farming that never existed pre-colonisation.[58] Water storage in dams and weirs has altered the natural flows of the river system since development began. These interventions have changed the volume, timing and duration of flows. According to the Murray–Darling Basin Authority, dams and weirs have prevented most small- to medium-sized floods that would have occurred in the past. The Murray–Darling Basin Authority also believes, based on evidence from the Wentworth Group of Concerned Scientists, that in the pre-development era, droughts occurred in the lower reaches of the system in 5 per cent of years. In contrast, in the post-development period, droughts occurred in an estimated 60 per cent of the years.[59] These statistics suggest that drought is directly associated with human interventions, but understanding which interventions were most significant is vital. And given the system's level of interventions, restoring certain environmentally sensitive areas is impossible without removing infrastructure like dams and weirs.

Several farmers I spoke with, including Darren De Bortoli and Louise Burge, argued that the natural ecology of the entire system has been radically altered and therefore cannot be restored through increased flow rates. At the end of the system, the Coorong, Lake Alexandrina and Lake Albert (the Lower Lakes) constitute a Ramsar site. During the Federation Drought, the South Australian government first discussed converting the Lower Lakes into freshwater storage. Louise Burge noted that that would have never been permitted on environmental grounds today. But by 1939, the project was completed, and barrages were built on Lake Alexandrina, separating the lake from the Coorong and the sea. They built 7.6 kilometres of concrete

---

58   Pascoe 2018.
59   Wentworth Group of Concerned Scientists 2017.

infrastructure, and the lake is only 1.5 to 2 metres deep.[60] To irrigate, they simply opened the floodgates at the side of the lakes, and the water would flow down to the farms. Louise explained that while the idea of building the barrages was received with enthusiasm in 1939, there was some apprehension in the scientific community about the potential effects, particularly that the construction of a weir or dam in the tidal compartment of a river would result in shawling (when the soil builds up around the mouth of a river) and sedimentation.[61]

The barrages were established in 1939, but the Murray mouth did not close until 1982. In those 42 years, the flows no longer went into Lake Alexandrina from the Coorong; they were funnelled towards the Murray mouth. Since the construction of the barrages, an elevated freshwater system in the Lower Lakes has been created. These elevated lakes cannot bring water in from the south-east of South Australia through the Coorong, which they had originally done. Due to the construction of the barrages, and the reversed flow of the Coorong, water can only come from the Murray–Darling system, which can never provide enough water in a drought situation. Given that the lakes cannot receive flows from the south-east and can only be fed through fresh water from the Murray–Darling Basin, there is now a black sludge accumulating in the Coorong because it cannot receive flows from the south-east. Human interventions indeed caused the environmental disaster in the Coorong, but what is less clear is how simply adding water will rectify the situation.

Before the barrages were built, the Coorong flowed into Lake Alexandrina and was fed by the lakes from the south. With the building of the barrages and farm drains emptying into the Coorong, there were no pushing flows from the ocean. The black decaying vegetation at the bottom of the system can never get flushed out to sea: "It just sits there festering like septic waste."[62] De Bortoli told me that due to the construction of the barrages and restriction of the saltwater flowing into the Coorong and Lower Lakes, the ecology at the end of the river has been completely changed. By 1982 there was not enough water

60 Murray–Darling Basin Authority 2023.
61 L. Burge, personal communication, 2016.
62 D. De Bortoli, personal communication 2016.

going to the Coorong so that some could flow into Lake Alexandrina, and by 1982 there was no water going into the Coorong. The point of real contention for many farmers is that the government will not acknowledge in its management plans that the Murray–Darling Basin is a human-made system – and that it is unsustainable to begin with.

One of the Commonwealth government's most important goals has been to keep the mouth of the Murray River open. One government representative told me that, according to geological records, in 1983, the mouth of the Murray closed over for the first time since the Ice Age. He said that since that occurred, a new river mouth punctured through the sand dunes, and 60 to 70 per cent of the time, constant dredging is needed to keep the mouth of the Murray open. This government representative told me that while there is a range of consequences if the mouth of the river is not kept open, the government's goal is predominantly to protect the whole Lower Lakes and Coorong, both listed as protected under the Ramsar Convention.[63] The government intends to protect those sites by putting more water down the Murray. But according to the farmers, and as described above, it is an impossible task so long as the barrages artificially elevate the Lower Lakes.

Two-thirds of the way down the Darling River, and upstream from the Coorong and Lower Lakes, is a whole series of natural lakes referred to as the Menindee Lakes system, and these "natural lakes" are being used as storage.[64] From 2013 to 2016, the section of river downstream of those lakes had essentially experienced low to no flows.[65] According to Johnson, the Murray–Darling Basin Authority is attempting to address the water needs of the Coorong and Lower Lakes areas by assessing water use upstream of those areas. From the perspective of farmers like Louise Burge, Darren De Bortoli, Helen Dalton, Derek Schoen and several others, this focus ignores the interventions in the Lower Lakes and Coorong as a potential source of the problems. This is a disciplinary effect of a discourse that privileges environmental water flows as the primary solution to the problem. Green environmental discourse, focusing on protecting natural sites, has had the disciplinary

---

63 Anonymous government official, personal communication, 2016.
64 H. Johnson, personal communication, 2016.
65 H. Johnson, personal communication, 2016.

effect of narrowing the government's focus. The farmers, on the other hand, challenge this narrow framing.

In much the same way climate cycles are debated in terms of natural cycles, the value of water flows in terms of restoring natural conditions is contested. The arguments on both sides rest on the validity of evidence gathered to indicate pre-development conditions, an approach in line with green environmental discourse. David Dreverman of the Murray–Darling Basin Authority said that the Lower Lakes were mainly fresh water and occasionally estuarine in periods of low river flow before the barrages were built. Dreverman told me it would be fresh water if they put 13,000 gigalitres on average through that system, although there might be saline intrusion around the channels where the lakes flow out into the sea. He added that there would be short periods when they would be estuarine, particularly in dry years. He believes the system was essentially a freshwater body, and the Murray–Darling Basin Authority has proven that by examining fossilised single-cell organisms (diatoms) found in sediment cores and tracing many of them to freshwater varieties.[66]

At the same time, Darren De Bortoli had been working with scientist Peter Gell to try to disprove many of the assumptions of the Murray–Darling Basin Authority, particularly the Murray–Darling Basin Authority's belief that the Lower Lakes were not estuarine. Gell is a paleoecologist at Federation University who examines changes in the conditions of wetlands over time. He has tried to advance a better understanding of "natural ecological character" as defined under the Ramsar Convention. He specialises in using diatoms as indicators of present and past conditions in rivers and lakes, particularly in coastal systems and across the Murray–Darling Basin. Gell made public statements in 2019, arguing that the Coorong and Lower Lakes had historically been mainly estuarine. In July 2019, the Murray–Darling Basin Authority responded to Gell, saying that removing the barrages would require even more water to be taken from production and that the barrages were initially built to maintain the balance between fresh and estuarine systems in the estuary:

---

66   D. Dreverman, personal communication, 2016.

The MDBA welcomes scientific inquiry and review, and we will consider the latest contribution by Professor Gell as part of that process. The Basin Plan is and continues to be founded on the best available science. It is based on a range of scientific reports, including an analysis of 114 years of the Basin's climatic conditions. The science is regularly reviewed by an independent scientific committee ... The Basin Plan is designed to achieve the best possible environmental outcomes across the whole Basin, not just the Lower Lakes and Coorong. The MDBA has consistently said that historic records from before the development of the river point to the Lower Lakes varying between being mainly fresh water with periods of estuarine conditions during periods of low river flow. The Lower Lakes cannot return to pre-development conditions because there is not enough water coming through the system to maintain the balance between fresh and estuarine. That's why the barrages were built in the 1940s. To return the Lower Lakes to their pre-development condition would require greater volumes of freshwater to reach the end of the system, and this could only be achieved if much less water was extracted upstream than required under the Basin Plan.[67]

After visiting the Coorong and Lower Lakes, I was shocked at the level of decay. A nauseating smell permeated the whole area, and virtually no birds lived there. It was also bone dry during the wettest year in over a hundred years. Dried-up salt lakes dotted the landscape, and blue-green algae were pervasive in the water remaining in the Coorong. Sedimentation is so high that there is stinky black sludge where water remains. But the Murray–Darling Basin Authority sees no option but to keep the lakes as freshwater lakes. The South Australian government argues that the lakes can never be salt water again. Louise Burge told me that the state government had said they could not open the barrages because there would not be enough fresh water to let the sea back out. The Murray–Darling Basin Authority, as evidenced in the above excerpt, has accepted the argument put forward by the South Australian government that it put the barrages in because of too much extraction

---

67   Murray–Darling Basin Authority 2019b.

upstream. But, according to farmers like Burge, the South Australian government made a conscious decision to turn the lakes into freshwater ones. Burge told me she does not argue for pulling the barrages down because she knows the political reality will not let that happen, but, on environmental grounds, she believes it should happen.[68]

Efforts to improve the conditions of the basin are affected by the belief that people can turn the system back to a previous condition. While the governments recognise that they cannot revert the system to pre-development conditions, they use strategies that could only work if the system was in its pre-development condition: that is, where there are no dams and the Coorong had normal flow patterns. Burge commented: "nature evolves, and nature is constantly changing, but we are trying to grab a moment in time of nature and say this is ideal or perfect".[69] The problem with trying to "grab a moment in time" and elevate it as an ideal is that there is simply no way to know what it was like. The effects of human development make it impossible to return nature to some ideal state. The green discourse presents a romantic view of nature as something that needs to be protected from human development but fails to recognise that we have been an essential part of nature that has irreversibly shaped the natural landscape in myriad ways. A critique of green environmentalism reveals that it is impossible to reverse the damage done to the environment or revert to some ideal point in time; it is only possible to imagine the social and environmental outcomes we collectively wish to achieve and work towards them.

Some of the fiercest resistance towards the Murray–Darling Basin Authority plan can be explained by the adverse impacts of environmental watering, particularly what are referred to as "blackwater events". Blackwater events occur when there is a lack of oxygen, resulting in mass fish die-offs. These events can occur when excess water causes trees and other debris to fill the water and cause the oxygen levels to drop, sometimes to the level of hypoxic (where there is not enough oxygen for fish or other animals or plants to survive). Many farmers are upset at what they see as poor environmental watering

---

68   L. Burge, personal communication, 2016.
69   L. Burge, personal communication, 2016.

practices, inundation as a result of environmental flows and an increase in toxic blackwater events. The high occurrence of blackwater events can be seen as a productive effect of the privileging of "environmental water" flows characteristic of green environmental discourse. The Basin-wide Environmental Watering Strategy (2014) stated the following:

> Assessments of tree stand conditions throughout the Basin in 2013 and 2017 have shown how floodplain forests and woodlands have responded to environmental conditions and management over time. River red gum forests and woodlands that line waterways and low-lying areas have generally responded well to management, while the condition of communities higher up on the floodplain (such as black box woodlands) or where constraints limit flows continues to decline.

Under the heading "Vegetation" in the draft plan, there are no in-text citations that refer to specific studies and research used to determine the effects of watering on vegetation. There are five references at the end of the report, none of which deals with river red gums or other native trees. The rest of the citations are from other Murray–Darling Basin Authority reports.[70] It is unclear what information is relied on to decide how trees should be watered.

Some government representatives believed farmers found it difficult to support the plan because they could not observe the ecological outcomes. Hilary Johnson, for instance, believes that farming communities struggle to see fundamental differences in the landscape with or without environmental flows because the changes are probably incremental or might not be obvious. He told me it is difficult for farmers to understand the benefits when they "might be hidden away in some wetlands far away from farms".[71] Conversely, for at least some farmers, the problems with watering have arisen from a lack of "common knowledge" about the landscape in the Murray–Darling Basin Authority.[72] Given the government's claims that trees along the

---

70   Basin-wide Environmental Watering Strategy 2014.
71   H. Johnson, personal communication, 2016.

rivers are suffering because of drought, Buller did an independent investigation on the water requirements of trees in the region. She told me that the eucalypts, including the river red gums, are flood tolerant, not flood dependent. Therefore, if they are overwatered, they will die, but they are also highly drought tolerant and can survive for several years without water.[73] According to Helen Dalton, native plants in the region can withstand inundation and periods of dryness of up to ten years. These plants have adapted to the Australian climate in this way.[74] Farmers believe that governments simply do not comprehend the requirements of the trees and land in the region.

At the time of our interview, several blackwater events had become hypoxic. Farmer Shelley Scoullar told me that, generally, blackwater events are okay in terms of occurring regularly and without critical impacts. However, Scoullar said that older farmers have expressed to her that while blackwater events were common in the past, it was not until 2010 that they started becoming hypoxic. Older farmers believe that the watering of forests results in more tree litter on the floor. When floods occur, the water becomes saturated with nutrients from the trees, and oxygen levels drop very quickly.[75] Farmer Bernard Walsh also reported seeing trees falling into the rivers because too much water goes down the river, and areas that should be dry are being flooded yearly. Walsh said governments seem unaware that the trees are not supposed to be flooded every year. Further, he told me that some areas should only flood every ten years, but now they are being inundated yearly. If the trees do not dry out, they will become waterlogged and die. He reported becoming aware of multiple incidents whereby the Commonwealth government flooded fresh water into a lagoon, and all the fish died due to the water becoming hypoxic.[76]

Farmers point to the increasing regularity of blackwater events as proof that the science around the basin plan is failing. The farmers hope to undermine the dominant narrative that the volume of water

---

72   H. Dalton, personal communication, 2016.
73   D. Buller, personal communication, 2016.
74   H. Dalton, personal communication, 2016.
75   S. Scoullar, personal communication, 2016.
76   B. Walsh, personal communication, 2016.

is the cause of problems in the basin. Farmers challenge the dominant discourse by showing that governments' desired environmental outcomes cannot necessarily be achieved in highly modified environments. Environmental watering appears to be based on assumptions about the requirements of the trees in the Murray–Darling Basin but, at least according to farmers and local witnesses, watering may be having severe detrimental effects on trees and aquatic life surrounding the forests. There is an assumption that putting more water down the river will restore the system, but such an approach can not only flood farmlands but also have negative consequences for the trees and aquatic life in the rivers.

Green environmental discourse has significant consequences for rice farmers. To outside observers, rice farming appears to require unsustainable amounts of water. Farmers insist that rice can be a vital aspect of a sustainable farming system and provide numerous environmental benefits, including the capacity to store and reuse water and provide habitats for species of wild birds. Farmers challenge the view that rice farming is destructive by trying to demonstrate that rice farms can reflect many of the same characteristics as "natural" spaces.

Most Australian rice farms are in the Murrumbidgee and Murray regions of the southern Murray–Darling Basin. The central irrigation districts of Murrumbidgee, Coleambally and the Murray Valley have suitable clay-based soils for rice growing on relatively flat land with established irrigation infrastructure. The centres of rice growing are around Leeton, Griffith, Deniliquin and Coleambally. The Rice Marketing Board of New South Wales legally owns all rice grown in the state under the *Rice Marketing Act* of 1983. The Rice Marketing Board distributes licences to approved buyers who purchase rice from growers.[77] Rice growing depends entirely on irrigation water. If there is a low water allocation, little to no rice is grown. Rice can only be grown on approved soils and is closely regulated by the water-use policies of irrigation corporations. The Australian rice industry is the world's leader in water-use efficiency, using 50 per cent less water than the global average. According to the Commonwealth Department of Agriculture, Water and the Environment, "Water use per hectare

---

77   Ashton and Van Dijk 2016.

continues to decline because of the industry's commitment to developing high-yielding rice varieties that use less water, and the use of world's best management practices."[78] Despite the relative success of rice farming in Australia, there is a debate over its sustainability. Critics have questioned whether farmers should be growing rice. Farmers, in turn, have responded by trying to reframe the debate and characterise themselves as environmental caretakers. In reframing the debate, farmers hope to garner support for their industry and gain recognition for their efforts towards sustainability.

Russell James of the Murray–Darling Basin Authority told me that, in broad terms, if water is taken out of the river system and used to grow a crop, the question of which crop is being grown does not matter to the government. But James believed that some agricultural production systems are more environmentally friendly than others. For instance, he claims that the environmental impact of growing other types of plants, like fruit and vegetables, is lower than rice.[79] Therefore, even though the government's official position is that they are concerned with how much water is being used overall, there are biases in thinking about which crops are more viable.

Farmers seek to challenge what they see as misunderstandings regarding people's knowledge of rice farming. Farmer Debbie Buller explained that when people fly over rice-growing areas in September, they will see what looks like a giant lake because the rice has just been planted. They do not understand that the water has just gone on the ground, that the water is very shallow and will be held for a long time in clay soil to grow the rice.[80] Fields are flooded at the beginning of the crop and the water sits there for the duration of the plant's growth as the plant slowly absorbs it.[81] Nonetheless, the popular press has drawn negative attention to rice growing in Australia in headlines like "Stop growing 'thirsty' rice: expert" and "Should cotton and rice be grown in Australia?"[82] Helen Dalton explained that farmers are never made to

---

78 Australian Government Department of Agriculture, Water and the Environment 2015.
79 R. James, personal communication, 2016.
80 D. Buller, personal communication, 2016.
81 H. Dalton, personal communication, 2016.

grow rice, but they choose to grow rice because it is the most efficient use of water. For Dalton, not only is rice efficient to grow, but it also adds value through the processing sector. Rice growing allows farmers to use any residual water left in the soil to grow winter cereal crops like oats.[83]

Farmer Debbie Buller remarked on the efficiency of growing rice in certain regions in Australia. "Australian irrigators are right at the top of their game. We've been rice farming for over 30 years, and we now grow twice as much rice with half as much water."[84] Further, rice farming has provided valuable habitat for bird populations. Helen Dalton told me native birds are abundant on her farm and throughout the area.[85] Farmer John Bradford spotted a pair of male and female endangered bitterns living and breeding in his rice fields, so he contacted Birds Australia to report the sighting. Bradford saw the occasion as an opportunity to demonstrate the environmental benefits of growing rice and to counter claims by green activists that rice is environmentally damaging. Birds Australia came to his farm within a couple of days after he made his report. He then informed the Ricegrowers' Association, and they have also been tracking the bitterns in the rice.[86] Bradford claims that the Ricegrowers' Association now has people worldwide inquiring about donating money to bittern programs. While there are many bird species on farms, the farms are not considered natural environments.

Despite the claims made by farmers, the dominant green discourse perceives rice farming as disadvantageous to environmental outcomes. Hilary Johnson from the Murray–Darling Basin Authority responded to farmers' claims by saying that productive lands are not gaining the same kind of biodiversity as would be exhibited in a wetland habitat. He does not deny farmers' claims about Australasian bitterns. Still, as a point of comparison, he told me that when the government inundated the Barmah–Millewa Forest, it was estimated that 20 per cent of the

---

82   Alexander 2019; Kennard 2007.
83   H. Dalton, personal communication, 2016.
84   D. Buller, personal communication, 2016.
85   H. Dalton, personal communication, 2016.
86   J. Bradford, personal communication, 2016.

known population of Australasian bitterns in the world turned up to that forest. Though he admitted it was hard to monitor, they were thought to have bred during that time.[87] Johnson believes that farmlands do not have the complementary habitat needed for the long-term maintenance of the bitterns. Nonetheless, it is evident that in some cases, farm habitats effectively provide refuge to endangered bird species, which could provide an opportunity for governments to preserve these species. This is the conclusion important to the farmers I interviewed.

In response to concerns about the potential value of farms as sites for ecological preservation, the government has begun to engage with farmers to look at some of these spaces and has introduced programs to help facilitate their revegetation. Hilary Johnson of the Murray–Darling Basin Authority told me that there are efforts on the part of the government to realise the potential environmental values of farms. The government is working with local communities and has entered into an agreement with Renmark Irrigation Trust in South Australia. Renmark Irrigation Trust is maintaining several sites with river red gum forest or black box forest that could be improved. Environmental water will be delivered through the trust's irrigation system to environmental sites within the agricultural community. Though involving small amounts of water (1 or 2 gigalitres), the government is seeing if the model can be applied in other irrigation communities. These amounts, of course, are much smaller than the volumes they are looking to deliver in the Lower Lakes and Coorong, where some 700 to 800 gigalitres of water a year are expected to be delivered. Johnson told me that these smaller community-based initiatives also support farm community engagement in delivering water to the high-priority zones. These smaller projects are important because farmers can generally maintain pockets of vegetation by flushing some of their excess water into these areas. The government wants to be able to help them provide additional water, so they can keep those pockets of native vegetation and wetlands that sit within the farming district in a healthy state.[88] While these programs are still in their infancy, governments are beginning to

---

87 H. Johnson, personal communication, 2016.
88 H. Johnson, personal communication, 2016.

recognise the importance of preserving parts of farms as integral to their broader environmental goals.

Farmer Helen Dalton told me that the government has failed to understand the potential positive implications of recognising how pristine environments and productive areas can work together: "'Beneficial' means not just productive, it means beneficial for the little bugs in the soil, beneficial for the critters on the crop, birds, frogs, etc. Beneficial for the environment and healthy communities."[89] Given the wildlife species on her farm, Dalton sees her farm as both a productive and natural space.

Farmers are paying high premiums for their water and, as such, will take every measure to see that water for irrigation purposes is retained on their farm and sequestered by crops, not flushed down the river. While there are variations in the initiatives and advancements undertaken by individual farmers, many understand the value of conservatively using water and ensuring that the river and surrounding wetlands are not contaminated by waste. As we see from this discussion, farmers seek to challenge the green environmental discourse by questioning the separation of productive and environmental water. While the green environmental discourse has focused on environmental flows, farmers seek to establish their role in managing water on farms and protecting environmental assets.

Some government representatives are beginning to recognise some problematic assumptions embedded in environmental discourse, particularly about the separation of "natural" and productive spaces. This is important because policy decisions are often based on defining the problem as dichotomous. Policy choices tend to reflect a view that decisions benefiting environmental spaces are detrimental to productive spaces and vice versa. This perception appears to be changing. For example, Hilary Johnson of the Murray–Darling Basin Authority admitted that even he struggles with the notion of a natural system. He explained: "there is no doubt that the system, by and large, is not natural" and that the Murray–Darling Basin Authority "manage[s] for environmental values". These values, according to Johnson, are determined through international conventions and the choice of

---

89   H. Dalton, personal communication, 2016.

governments to list certain areas as Ramsar-protected sites. Johnson told me that while they do not consider all these sites to be natural, Ramsar recognises that even sites that are entirely human-made have significant environmental values attached to them. The notion that something might be less natural and therefore has no environmental value is something that he struggles with.[90] In essence, both farmers and Johnson agree that an environment's status as natural should not determine its environmental value. However, by this same logic, a farm should also be assessed by its contributions to environmental values, even if it is an environment affected by human development. Johnson's is a powerful admission that some government employees understand that environmental policy often stems from socially constructed values concerning the environment. These constructions can have profound effects on both environments and people.

Johnson expanded on his earlier point, saying that judgements must be based on people's environmental values. Further, in terms of protecting and restoring environments, there will always be limitations within a system that is comprised of numerous dams, weirs and locks. As discussed in previous chapters, all this infrastructure has fundamentally altered the water delivery patterns in the system and the volumes of water available.[91] This reality, which farmer Ian Mason also articulated, is that only some two hundred years previously, none of these European-style farm communities would have been able to survive in the natural environment.[92]

Such comments by government representatives (and farmers) appear to represent a split in the green environmental discourse. Opportunities are emerging to find common ground among farmers and government officials, thereby creating the space needed for a new discursive response and different policy solutions. With this recognition of the problems associated with trying to determine what is a natural system, there is also an important acknowledgment that the basis for much environmental decision-making lies in the kinds of values that people hold dearest. The next chapter discusses that

---

90 H. Johnson, personal communication, 2016.
91 H. Johnson, personal communication, 2016.
92 I. Mason, personal communication, 2016.

Table 4.1 Green environmentalism: transcripts and policy, legislation and actions in the Murray–Darling Basin

Transcripts, metaphors and rhetorical devices	Policy, legislation and actions taken
protection	*Water Act* (2007)
wilderness	Australian Greens (significant after the 2001 election)
natural versus unnatural	
"land … of droughts and flooding rains"	Ramsar Convention (1975)
	Basin-wide Environmental Watering Strategy (2014)
environmental protectors	
environmental values	
environmental water	
natural weather cycles	
nature	

community-centred values may be fundamental to positive environmental outcomes.

Table 4.1 summarises some of the transcripts that regularly appear in the green environmental discourse, as well as policies, legislation and actions taken.

Conclusion

Farmers resist the discourse of green environmentalism in several ways. First, they challenge the science of water cycles that underpins Murray–Darling Basin management plans by drawing on alternate scientific explanations and local knowledge (including their own community's environmental knowledge) to say that drought is "normal" in Australia and that the government is misinterpreting the cycle of drought. They also challenge the basic premises of the management plan from within the plan's logic: if the cycles are less predictable, then the plans based on these cycles are also likely flawed. Second, they try to show how critical environmental issues (such as survival of the lower lakes) cannot be reduced to a question of inadequate amounts of water, nor can these issues be solved by simply "adding more water". Once again, using alternative knowledge produced by scientists as their justification, as well as evidence from their knowledge of their farms, farmers are working to redefine the problems in the Murray–Darling Basin in ways that reveal how issues are about much more than simply increasing the flow of water to "nature". Third, as a key challenge to environmental water, farmers are mobilising evidence and interacting with actors like Australian bird conservation organisations to show that productive farms can also be healthy ecosystems that harbour important species. These arguments challenge the basic dichotomy between nature and farms and thus contest the assumptions embedded in green environmental discourse. Finally, farmers seek to present themselves as environmentalists in the Murray–Darling Basin and reposition themselves within the debate. Through examples that show they are introducing efficiencies and providing habitat on their farms, they are trying to shift the public and governments' perception of them and change the environmental debate in Australia. Further, interviews with government officials revealed that the farmers are having some effect in their efforts to reframe the debate. Officials recognise that farms can and do provide wildlife habitat. Still, the evidence shows a disconnect and lack of trust between these two groups, with the farmer perspective making minimal inroads into public perception or government policy.

Green environmentalism encompasses a view of the world wherein human systems are generally regarded as detrimental to the natural world. Several scholars of the Murray–Darling Basin have focused their research on the negative environmental impacts of farming in the Murray–Darling Basin and what measures governments should take to redirect water from productive to environmental spaces.[93] This characterisation of the problem frames people and productive spaces as separate from nature, although many scholars have questioned such assumptions about "nature".[94] These scholars have argued that nature is inseparable from human communities and farmers also see the natural

---

93 Grafton and Horne 2014; Pollino, Hart, et al. 2021; Ross, Buchy and Proctor 2002.
94 See for example Cronon 1996; McKibben 1989; Oelschlager 1991; Soper 1995; Vogel 2015.

world as not external to the social world. As Bookchin[95] pointed out, framing humans as outside nature can easily lead to the subordination of human social problems to ecological problems while failing to recognise the potentially positive influence people can have on their environments. As we have seen in this chapter, the development of the Murray–Darling Basin Plan has been affected by a conceptualisation of the environment as separate from people. This view has its roots in the historical trajectory of the green movement in Australia, with its emphasis on romanticism and biocentrism. Farmers share many of the same environmental concerns as people living in cities (and government bureaucrats), but they define problems differently. The unique situation of farmers means they conceptualise environmental spaces differently. For instance, farmers tend to view nature as something that needs to be carefully managed by people living and working on the land. As well as contrasting with some aspects of green environmental discourse, this view challenges administrative rationalism wherein governments and experts are best positioned to manage the land.

As explained in this chapter, farmers reveal how people have always had an integral role in directly managing environments in their day-to-day activities, including both "productive" and "unproductive" spaces. Australia's green movement often fails to account for the significant impact that human societies have had on natural landscapes throughout history. Farmers seek to undermine the notion that productive and natural spaces are inherently separate. The examples provided in this chapter, including contested ideas about climate cycles, the causes of problems in the Lower Lakes and Coorong, and perceptions of the environmental impacts of rice farming, challenge the dominance of green environmental discourse.

Human interference has made it impossible to revert to some ideal moment wherein the ecology was "natural". Ultimately, decision-making is guided by the imaginations of those invested in the ecology of a given space and the values that these actors hold. It follows that if the goals of the Murray–Darling Basin Authority plan are to be based on shared values, then we must decide what values, and whose

---

95   Bookchin 1994.

values? As this chapter demonstrates, values are based on fundamental assumptions about the nature of the environment and human roles within it. The next chapter reframes all of this by presenting an alternative discourse that foregrounds the role of social relationships in determining environmental goals and outcomes. Suppose environmental values are ultimately determined by the values of actors within a given community. In that case, policymakers may benefit from turning their attention to social relationships and community values to make better human and natural environmental decisions.

5
Community-centrism

Community-centrism is an alternative discourse of resistance that seeks to put human social relationships at the heart of environmental decision-making. Community-centrism provides a reconceptualisation of environmental problem-solving in the Murray–Darling Basin, surfacing economic, environmental and social opportunities. Farmers offer the basis for a unique approach to environmental problems that situate people and the social interactions between them as central. Community-centrism has been developed through dialogues with farmers and their unique perspectives on environmental concerns. The discourse can be seen as a challenge to the four environmental discourses discussed in Chapters 3 and 4. First, the discourse rejects the centralised planning prevalent in administrative rationalism. Instead, it envisions a bottom-up, community-based planning process. Second, the discourse adheres to many assumptions underlying economic rationalism but, in place of centralised economic planning, encompasses a vision of a more community-oriented and localised economic planning approach. Third, the discourse challenges governmental institutions and structures that limit social engagement and impede participatory democracy. The discourse thus challenges the mechanisms of democratic pragmatism that do not go far enough to ensure deliberative engagement in decision-making. Lastly, the discourse highlights the vital role of people as part of effective environmental management. This contrasts with green environmentalism, which focuses on protecting "natural" spaces from human activity.

Community-centrism entails a specific public philosophy with distinct problem definitions and policy solutions. Social relationships, community health and community cohesion are central ontological values. Economic and environmental outcomes are fundamental to the discourse but are directly tied to social relationships. From an epistemological perspective, strong social networks allow knowledge transfer to occur. While scientific and academic knowledge is important, collaborative scientific research and experiential knowledge also play an essential role. Further, the discourse recognises that the value of knowledge depends on the social capital of the knowledge holder. In community-centrism, knowledge holders who are directly involved in the community and invest in those communities have greater social capital. The public philosophy of community-centrism also gives rise to certain problem definitions. Problems are articulated and interpreted by the community, and solutions are generated through processes of community engagement. The discourse highlights the significance of local agents as environmental stewards, the importance of knowledge sharing within communities and a bottom-up approach to problem-solving. The solutions to address water issues in the Murray–Darling Basin that stem from this approach include community-based monitoring and reporting initiatives, strategic community-led decisions on water buybacks and farm-led environmental programs.

Variations of community-centrism appeared among the farmers interviewed. To some degree, all the farmers interviewed expressed that they depended on their communities to run their businesses effectively. Some farmers highlighted the significance of knowledge sharing among their peers, others explained the importance of the local towns in supporting their operations and others were engaged in community-based environmental initiatives. Farmers are not a uniform group, but all respondents shared a commitment to their communities and acknowledged that community-based planning and knowledge sharing contribute to environmental and economic outcomes. These shared values, as articulated through my interviews, are labelled as the discourse of community-centrism.

The philosophy of community-centrism

A community-centred discourse prioritises relationships and community cohesion. It also emphasises individual and collective health as a primary concern. The central preoccupation of the farmers interviewed, and the central ontology of the discourse, is the preservation of established communities. First, farmers are fighting to preserve the health of their families and their way of life. They are concerned with keeping services like schools, shops and secondary industries in their communities. Farmers prioritise maintaining community structure above all else. Second, there is a focus on the reliance of farmers on those around them, including families, other farmers and workers. Interviews suggested that focusing on economics and the environment in the abstract – divorced from a sense that community lies at the heart of such economic and environmental considerations – can cause people to overlook the societal consequences of reform. Farmers in isolated communities rely heavily on their families, other farmers who surround them and their employees for the emotional support necessary to live a "normal" life. Women are central to this community support structure. Given this ontological orientation, farmers are concerned with intergenerational continuity and the impacts of drought on the long-term prospects of family farming.

Drought and water reform have had an enormous impact on the social lives of farmers, a problem that has not received much attention. Farmers noted just how dependent their communities are on surrounding towns, and that these towns also rely on them. The drive towards larger, more efficient farms has contributed to the breakdown of local rural life, and drought and government policy have often made conditions worse. Some towns struggled before the Millennium Drought, but despite farmers' knowledge of the climate cycle and the recurrent nature of drought, its extreme duration and severity caused such a hit that they collapsed. Derek Schoen recalled that at one point during the drought, there were 250 houses for sale in Deniliquin and no buyers.[1] As a discourse of resistance, community-centrism seeks to reorient the policy agenda towards rural communities as the primary

---

1   D. Schoen, personal communication, 2016.

focus. Interviews suggest that a central concern of farmers was not just the viability of their farms but of the larger communities they support. Outcomes like the closure of schools and the breakdown of community and social structures (sometimes resulting in suicide, high incidences of divorce and drug addiction) appeared to be the primary concern of farmers in these communities.

The social relationships between farmers and communities are critical in terms of maintaining the quality of rural life, environmental sustainability and the productive capacity of businesses. Illustrative of these dynamics, Barry Kirkup explained that farmers historically enjoyed the support of their local governments and communities. Communities supported farmers because they understood that taking water out of the system would result in economic consequences for local businesses like schools, hairdressers and tyre businesses. The farmers received support from local communities, particularly in Leeton, where some 600 community members are employed by just one company, SunRice. Farming allows other industries to remain viable, including nut processors, wineries and cattle feedlots.[2] This community-based support often encouraged farmers to continue farming and provided them with essential resources to help them survive during the most challenging periods. For many of the farmers interviewed, farming is seen as a social calling, and they want to be acknowledged for contributing to the larger community, not just to the economy. The centrality of community to this discourse was also evident when farmers spoke about how the maintenance of the community was missing from policy responses to drought and water reform. Farmer Barry Kirkup noted that one impact of water reform is that the costs of services have continued to increase for those who are left trying to maintain the farming systems. Fewer people remain in the system, but infrastructure costs remain the same. Kirkup said that all three of the farms next to him were sold at the same time, and the channel where Kirkup gets his domestic water (3 megalitres a year) was closed. The irrigation company would not let water into the channel just for 3 megalitres because it takes about 10 megalitres to fill the channel. The domestic water channel was classified as a stranded asset.[3]

---

2   B. Kirkup, personal communication, 2016.

This type of situation means people living on farms have no drinking water and instead must purchase it for daily domestic use. Families with children found it particularly challenging to live in such conditions. Many abandoned their farms to move to towns and cities or left their husbands and fathers to manage alone.

The interconnections between farms and communities allow farms to endure. A community-centred approach acknowledges that the impacts of reform are not just confined to the farm. There are knock-on effects throughout communities, diminishing the capacity of entire rural towns to remain viable. During drought periods, farmers still paid water charges even though they had no water. Farmer Helen Dalton recalled how they budgeted during the drought by not going to the produce store or the mechanic, which ultimately caused significant damage to the larger community. Farmers needed the local businesses to stay in the towns. The farmers understood that their continued existence depended on ensuring that the townspeople had the financial support they needed. They knew that if local businesses in the community left, then they would not be able to continue farming. Helen Dalton said she has an aversion to talking about "farmers". Instead, she prefers to talk about communities because farmers are integral to the larger community. She observed: "People think farmers 'whinge', but it's about all of our communities. It's not just about being able to produce something; it's about supporting our community, and them supporting us."[4] Community-centred discourse reveals that farmers cannot be viewed in isolation from the larger communities in which they live.

The discourse also demonstrates the importance of personal relationships, such as with farm employees. Farmers depend on labour that is only available if communities can receive and accommodate workers. Further, providing consistent work to labourers is the only way to keep them in these communities. For instance, during the drought, Helen Dalton and her husband had to let go of all but one permanent employee. Since the drought, they have been able to employ more people, but she noted that many farm workers left the area during the

---

3   B. Kirkup, personal communication, 2016.
4   H. Dalton, personal communication, 2016.

drought, so it became more difficult to find qualified workers.[5] When labourers leave the communities, getting them to come back is difficult. This is particularly the case with skilled labourers. Community-centrism highlights the value of people like farm workers who provide the vital support farmers need to retain their businesses.

Community-centrism also pays close attention to and underscores women's critical roles in sustaining vibrant and healthy communities. While women farmers paid closer attention to gender dynamics, both men and women interviewed discussed how central women and mothers were in maintaining farm operations. One of the most significant social changes resulting from the drought and the exodus of people from rural communities was workforce restructuring. This restructuring affected women's roles, which reverberated throughout the broader community. Many women, like Debbie Buller, went away to work as nurses or teachers; and some started their own businesses. This exodus meant that men were left to tend to the farms on their own, and the relative isolation led to severe impacts on mental health.[6] Buller and Dalton told me that farming generally requires a more traditional family structure. Historically, mothers did not work outside the home as they are tasked with caring for the children and the husband. Helen Dalton, for instance, said she had always taken on the role of a traditional farm wife and mainly devoted her life to caring for her kids. Dalton explained that many women who left the farms to work never returned to their husbands. Dalton's brother sold the farm after his wife left to work. He was too lonely on the farm to continue on his own.[7] For many women, the change meant a better life with successful businesses or careers. But for the men, it often meant they could not continue farming because the lack of connection and support was just too debilitating.

Drought and water reform drastically reshaped the structures of the community. These changes were of deep concern for the women who remained. In the town of Griffith, the drought dramatically changed women's roles in the community. During the drought, some

---

5   H. Dalton, personal communication, 2016.
6   D. Buller, personal communication, 2016.
7   H. Dalton, personal communication, 2016.

of the women interviewed met for Thursday lunches to discuss their various activities. They were getting jobs, writing books and doing other things to keep their minds occupied, and earn extra income. The men stayed on the farms to look after the stock and to make cuts to stock: "The most horrible job, I think, is feeding stock when they are hungry and starving, trying to keep them alive every day and carting water. It is the most soul-destroying job." Dalton recalled how difficult it was for the men, year after year, to cart water to thirsty animals as the drought continued unabated. Most of the women are still not back on farms. Dalton was the only woman left in farming in her small community. The impact on the community was far-reaching; no one had time to help with the work done by the local church hall or to go to bushfire meetings. She lamented: "all these important things that kept the glue of the community together have virtually disappeared".[8] Community-centrism brings attention to how environmental issues penetrate the whole community.

Community-centrism also highlights how healthy farm communities depend on multigenerational continuity. As the older farmers age and the younger farmers observe the hardships and increasing uncertainty involved in modern farming, young people leave the farms in search of alternative prospects. Barry Kirkup's son, for instance, had worked on the farm for a time but then opened an electronics business in Leeton. From Kirkup's point of view, the younger generation does not want to wait for their returns. Farming, he explained, is about long-term gains: "you are not in it for five minutes". Setting up an effective farming system requires substantial assets, and some years are lean while others are good. He believes the young people do not want to hang in there through the lean years.[9] In a similar vein, farmer Bernard Walsh said that his work had been made more difficult because he pushed all his kids off the farm at the beginning of the drought. After all, there was no work for them. He encouraged his children to enter the trades. But he now believed that if he had someone to sit down with and share his problems, he would be better able to handle the pressures: "I don't know if I'm being super negative, but I

---

8   H. Dalton, personal communication, 2016.
9   B. Kirkup, personal communication, 2016.

can't see a light at the end of the tunnel." Walsh explained that none of his children will return home to the farm because the risks are simply too high. He said that while he loves farming and has no problems dealing with the associated tasks, coping with the water situation, "which is the lifeblood" of their business, was too difficult in the current climate. Social isolation and increasing pressures make farming too difficult. He constantly worried about how much allocation he would receive because of the heavy dependence on the water market. With these conditions, he always "has to be thinking about marketing and not farming".[10]

Losing so many people has been devastating to those left behind. And it is not just farmers who leave but everyone who supports farm communities, like labourers, teachers, hairdressers and grocers. When there are fewer people in the communities, prices must be higher to have goods sent to the communities. As demand decreases, the price of everything also increases, so farmers are forced to find ways to make more profit.[11] Farming depends on the long-term participation of everyone in the community: men, women, grandparents, young people, small business owners, and community members like teachers and nurses. Farmers do not exist in isolation from the rest of the community; they are an integral part of and depend on the broader community that supports them. As we can see from these examples, the ontological orientation illustrated through these stories is one centred on community.

Community-centrism can also be defined by an epistemology that privileges local community-based knowledge, particularly farmer knowledge. While the discourse gives attention to scientific and academic knowledge, it favours collaborative scientific research and highlights the critical role that experiential knowledge can play. Fostering social networks for knowledge exchange is another key aspect of community-centrism. This discourse also holds that the value of knowledge depends on the social capital of the knowledge holder. Within community-centrism, knowledge holders directly involved in the community and who have vested interests in those communities

---

10 B. Walsh, personal communication, 2016.
11 A. Clark, personal communication, 2016.

are seen to have greater social capital. These are the people whom the community relies on to influence outcomes. While there is space for farmers to engage with government experts, under the current circumstances that knowledge is often viewed with scepticism and distrust. One of the insights I deduced from these characteristics of the epistemology of community-centrism is that if farmers are not given opportunities to shape the governance processes they are subject to, they will resist these processes. This insight suggests how important it is for governments and external "experts" to build trust through close engagement with farming communities. This epistemological approach represents a challenge to the discourse of democratic pragmatism that was explored in Chapter 3. While democratic pragmatism emphasises the practical application of ideas through democratic processes such as environmental consultations and consensus-building initiatives, the discourse often ignores the wider social processes in which specific environmental issues are embedded. By focusing on social processes, community-centrism offers a way of redefining problems that accounts for uneven power dynamics and how some types of knowledge may be privileged over others.

The epistemology of community-centrism is evident in the critique farmers have of current community engagement efforts by governments. Community-centrism represents a discourse of resistance to the dominant discourses that tend to privilege government and expert knowledge. Community-centrism is characterised by a focus on increasing social trust through empowering people. Farmer Helen Dalton believed that a lack of trust is the main obstacle to incorporating farmers in the decision-making process. Dalton would have liked to see a more "bottom-up" approach on the part of the government. She believed that, given the current "top-down" attitude, "the farmers will fight the whole way". As a primary school teacher, she offered an analogy based on her own experience:

> In the first year, you instruct the children in the basic rules of conduct; in the second year, you tell the students that they are in charge of their environment and the teachers are there to help them. By the third year, you ask the students what rules they want to apply to their classroom. When the teacher asks the students

to make the rules, the rules are generally ten times more stringent than if the teacher had made or imposed the rules. The students want the rules and will willingly abide by the rules because they made the rules and agreed to them. In this case, the teacher no longer has to impose or enforce any rules.[12]

According to Dalton, this is how to create good governance because the students have learned what they must do and agree with the rules because they developed them. Dalton's analogy demonstrates the significance of individual responsibility and accountability towards environmental planning processes. These values demonstrate that social trust is a defining characteristic of community-centrism discourse. The discourse asserts that when the people subjected to governance feel empowered, they can make better decisions and often go above and beyond expectations. Farmer Shelley Scoullar believed that reform must "build from the grassroots because the people who have been living in the system for generations understand how the system works". She added: "And if you work with local knowledge, we call it localism, then you can design a plan that will meet environmental outcomes and social and economic outcomes."[13] Locally based decision-making provides the opportunity to engage meaningfully with farmers. A community-centred approach recognises the potential role that governments can play in helping adapt to challenging circumstances and how community members might influence the forms that this help could take.

Community-centrism highlights the value of local farmer knowledge in the policy planning process. Collaborative, locally embedded approaches to research and knowledge dissemination are central to the epistemology of community-centrism. In the case of research, investment planning and strategic development, farmers see themselves as playing key roles in defining objectives and identifying the limits and desirability of certain projects over others. The discourse focuses on the potential positive impacts of actively engaging farmers in consultations, developing models and designing policy. For example,

---

12   H. Dalton, personal communication, 2016.
13   S. Scoullar, personal communication, 2016.

farmer Louise Burge explained that while it is good that the government is looking at redesigning more realistic flow regimes, she also thinks that both the New South Wales and Commonwealth governments could do more to work with local communities. For Burge, this would mean "getting the consultants out of the picture – get the bureaucracy and the consultants out of the picture – and key people come into the community and ask farmers how they would like to design a program to work through the solutions". In this vein, Burge has written up a constraints sheet that includes ways of designing environmental flows that could work for the community and the government.[14]

Community-centrism is focused on the role of local knowledge. Burge reported that she faced a significant amount of adversity in confronting the problems with the plan. But she has spent much time compiling what she calls the "Murray messages", which, she told me, are "solution focused". Since she anticipated that the government's objectives in the plan are not likely to change, she has tried to make recommendations that are not too politically difficult. Her "messages" were endorsed by Murray Irrigation and then sent to the Murray–Darling Basin Authority. Murray Irrigation has also discussed amending the *Water Act*, particularly the Sustainable Diversion Adjustment Mechanism. Burge said that farmers are continuing to look for ways to secure social, ecological and economic outcomes in the Murray–Darling Basin and to find pathways to get governments to work more closely with local people.[15] Burge's comments reflect a desire on the part of farmers to be included in negotiations, even when their goals are not necessarily recognised. Burge regularly emphasised the desire for the government to include "local" people. Conversely, the discourse draws attention to how government knowledge can be more accessible to farmer communities. This discourse is grounded in the assumption that effective knowledge exchange creates opportunities for cooperation and can help foster productive long-term management of water in the basin. But the value of these exchanges largely depends on how the other party is viewed. Community-centrism focuses on

---

14   L. Burge, personal communication, 2016.
15   L. Burge, personal communication, 2016.

how locally based experiential knowledge can gain legitimacy and how that knowledge can be made accessible and quantifiable by government agents and farmers.

The epistemology of community-centrism reveals that the value of knowledge depends on the social capital of the knowledge holder. Farmers tend to value knowledge according to the strength of their relationships. While farmers often act independently, they are also inclined to evaluate the knowledge of close friends and associates as important. Knowledge from government sources is seen with sceptical distrust by farmers, as the latter do not believe that government officials are invested in outcomes important to them and their communities. Some farmers use consultancy services, but they tend to enjoy long-standing and strong connections with these consultants. The social relationship with a knowledge broker is of utmost importance for farmers.

As explained above, one of the underlying epistemological assumptions of community-centrism is the recognition of farmer knowledge. This approach contrasts with the types of engagement and consultation we observed through analysing democratic pragmatism. Community-based discourse envisions a more deliberative democratic engagement. Elinor Ostrom has argued that a sense of individual responsibility and accountability towards the process characterises good governance.[16] Communities often find stable and effective ways to define the boundaries of a common-pool resource, define the rules for its use and effectively enforce those rules through active engagement in the process. The effective management of a natural resource often requires "polycentric" systems of governance where everyone affected has a role in the process.[17] My analysis shows that the farmers I spoke with place "community", ontologically and epistemologically, at the centre of their understanding of the issues and challenges facing the Murray–Darling Basin. The discourse thus encourages new approaches to water management, as well as water-policy development and planning approaches grounded in ensuring farmers and their communities have seats at decision-making tables.

---

16 Ostrom 2012.
17 Ostrom 2012.

An alternative problem definition

Community-centrism is an alternative discursive framework for understanding and addressing environmental issues in the Murray–Darling Basin. It challenges the discourses explored in previous chapters. The interviews with farmers revealed a unique way of understanding environmental problems that focuses on community-based outcomes grounded in practical experience and decision-making from the bottom-up. The discourse, therefore, generates potential productive opportunities for alternative approaches to water reform.

In community-centrism, information gains value based on the quality of a farmer's social relationships with the knowledge holder. But, within the scientific community, it is often assumed that information has value independent of the social conditions in which it exists. What are we to make of these contrasting perceptions and the relationship between them? For the farmers I spoke with, social trust and local relevance are essential factors in determining which information is valued and which is disregarded. In this way, community-centrism can be seen as a challenge to the dominant discourse of administrative rationalism, which privileges the role of experts and takes a largely top-down approach to knowledge dissemination. Community-centrism privileges more contextual and collaborative knowledge generation, meaning problems are often defined as locally specific and collective. The epistemology values the co-creation of multiple forms and sources of knowledge. While scientific knowledge holds value, there is a recognition that evidence provided by farmers and others in the community also provides value. The examples in this section illustrate how community-centrism challenges administrative rationalism and the alternative problem definitions and policy solutions that arise from this challenge.

The two key sources of knowledge that inform farmers are scientific and farmer based. Interviews with farmers revealed a common perception that scientific inquiry did not pay enough attention to the locally specific context. Also, according to several farmers interviewed, scientific and government information is not presented in a form that is readily accessible to farmers. Where it is available, its applicability depends on local conditions. A common complaint among the farmers I interviewed was that the scientific studies funded by the government often assumed that the results achieved in one area were universally applicable. The results of these scientific studies can only be duplicated if the same soil and moisture conditions apply. For example, as discussed in Chapter 4, efficiencies in rice farming depend heavily on the amount of clay found in the soil. In community-centrist discourse, there is a recognition that the value of knowledge is spatially dependent, meaning local conditions and preferences influence their applicability and relevance. Given these criticisms, it is understandable that farmers rely predominantly on information received from their counterparts. This dynamic demonstrates that the social relationships among farmers are key factors determining the kinds of knowledge that they find helpful and ultimately decide to implement.

Interviews with farmers reveal how deeply they have come to rely on one another's experiential knowledge. Further, farmers are not receiving knowledge from the government in a way that is accessible to them. Community-centrism reveals the potential for authentic and meaningful reciprocal engagement between farmers and government officials. This orientation contrasts with administrative rationalism, which constitutes a largely top-down engagement model. Within farming communities, there is a wellspring of shared knowledge largely untapped by policymakers. In the case of water management, for example, farmers rely on each other to make informed choices about irrigation intervals. For instance, Gary and Margaret Knagge's neighbour Chris discovered that if the temperature goes above 36 degrees Celsius, he needs to shorten his irrigation intervals to just three days.[18] A strong culture of community-based knowledge transfer is built on strong social bonds. The Knagge's knowledge of appropriate irrigation intervals was gained through personal relationships like the one with their neighbour, but collaboration with government extension services could have given him a better understanding of what to do with forecast heat spikes. Encouraging and fostering strong social bonds between farmers and government representatives would present many more opportunities for knowledge exchange. In this way, community-

---

18   G. Knagge, personal communication, 2016.

centrism challenges the discourse of administrative rationalism and provides an alternative model of engagement.

Similarly, Gary, Margaret and Chris have also learned how much water plants need depending on their stage of growth and the ambient temperature. When it is very hot and close to harvest, farmers know that plants need water to reach their tips. Knowing when to switch from a seven-day watering cycle to a three-day watering cycle is essential, but that knowledge comes almost entirely from experience and shared local knowledge. Similarly, rice farmers want to steward their permanent water allocations carefully, so they must make accurate predictions based on the weather for that week. Making the right decisions about watering is vital for farmers, and there is little room for mistakes. For corn farmers, if temperatures go above 40 degrees Celsius, the crop can fail in just one day without water. Not only can the wrong decision cost farmers money, it can also mean water is not being used efficiently.

For farmers like Gary, Margaret and Chris, timing is essential knowledge. While some of that knowledge comes from government-funded sources, most comes from listening to other farmers and paying attention to what is happening around them.

> A lot of the knowledge has to come from personal experience because it's real-time, real environment, who made the biggest boo-boo. The farmers are always the ones who have to pay for mistakes, while the scientists and the bureaucrats will still get paid no matter what kind of information they provide. These others will all be there the next year to continue, but the farmer may not be.[19]

Fostering social bonds also means that information must be readily accessible and comprehensible. Farmers reported having a great deal of difficulty accessing and understanding the information that was available to them from outside sources. Farmers without a university education have difficulty benefiting from researchers' knowledge. It must, therefore, be broken down into information that the farmer can use.[20] Community-centrism foregrounds collaborative community-

---

19   G. Knagge, personal communication, 2016.

based planning; it asks farmers and government officials to examine their positionality. Government officials cannot expect to provide advice and services to farmers like they would provide information to other government officials and researchers. Building relationships means looking at how knowledge is shared and if that knowledge is truly accessible. In this way, community-centrism presents a direct challenge to the high modernist top-down orientation of administrative rationalism.

Research by the government can support new technologies and longer term projects that private interests cannot pay for. Publicly funded research supports the public interest. Focusing on the significant social outcomes of water policies makes it clear that investing in agricultural research and technological improvements provides wide social benefits. There is much to be gained from active government engagement in research and on-farm support. For instance, research can play a crucial role in advancing farmers' economic interests, the communities they support, and the environmental interests of farmers, governments and environmental organisations. But who should decide what to research and who should pay for these investments? Farmer Ian Mason believed that farmers had no problem paying for research they expected to profit from, but universities and research institutions should invest in the early stages of basic research. He would have liked to see the government get more involved with farming communities to see what kinds of measures were needed so they could design research that would provide valuable and practical solutions. For instance, weed control costs farmers the time and energy needed to remove weeds physically, which can affect production. Inventions using microwave technology and robotics have great potential in identifying and eliminating weeds without using environmentally harmful pesticides.[21] Further, in terms of developing rice varieties, scientists must do molecular work to determine which varieties perform best. There needs to be a greater drive towards employing future techniques and technologies, and some of the costs of innovation must be borne by the larger Australian community, or they

---

20   G. Knagge, personal communication, 2016.
21   I. Mason, personal communication, 2016.

simply will not be undertaken. But in line with community-centrism, farmers and other community members can play a critical role in determining which technologies will work best in locally specific contexts. A collaborative, locally based approach can assist in this endeavour.

Since the larger Australian community wants to see a more sustainable environment, there must be ways to create the tools that farmers need to build such an environment. Farmers tend to define the environmental solutions, on the farm and off, as a social good. This way of defining the issue is a departure from economic rationalism, which tends to allocate responsibility to individual farmers, and from administrative rationalism, which has focused on the role of governments in managing environments. In administrative rationalism, knowledge production seeks to employ the best available science and expertise but is not guided and inspired by farmers' knowledge. Community-centrism recognises the importance of this interchange of knowledge between farmers, experts, scientists, government, local community groups, and the people in the communities.

Farmers provide an alternative vision for government involvement, beginning with community-based planning. In addition, the discourse reveals the limitations of administrative rationalism as a discourse. Community-centrism defines this problem as one relating to a deficiency in social trust and a lack of appreciation and understanding for the work of farmers. If policy interventions were focused on this problem of social cohesion and trust, it would be easier for governments to make significant inroads into farming communities to produce and disseminate information. Community-centrism provides an alternative lens to administrative rationalism, one that accounts for the contributions of the wider community, including farmers, government experts and environmentalists.

As discussed in Chapter 3, the farming community has gradually moved towards a greater concern for environmental outcomes and away from purely economic rationalist thinking. But farmers have struggled to differentiate themselves from other industries in how they give back to the broader communities they support. As was demonstrated in Chapter 3 in the discussion of economic rationalism, while some industries can pay more for water, that does not necessarily translate into more efficient or sustainable water use. Further, a free market approach often fails to account for social values. Community-centrism draws attention to the interconnected nature of social, economic and environmental outcomes. The discourse challenges many assumptions embedded in economic rationalism by focusing on social outcomes. A focus on community-based outcomes expands our perceptions of value and challenges the notion that outcomes can be measured purely by economic gains (often only short term). Under the discourse, economic problems account for larger societal values and longer term economic gains to the broader community. The discourse can therefore be seen as a direct challenge to economic rationalism and market-based instruments that simply make water a commodity that any industry or user can access. In what follows, I explain how community-centrism offers a discourse of resistance to how farm economies are currently understood. First, the discourse highlights the social and environmental values of farming, as opposed to extractive industries. Second, the discourse highlights the value of farming in terms of maintaining strong locally based economies. Other industries like mining and gas extraction tend to benefit export, trade maximisation–focused sectors that make minimal investments in secondary processing within Australia. Finally, farming in Australia provides numerous benefits to the broader communities. A strong agrarian economy supports tourism, provides incentives for environmental management, maintains vibrant and healthy rural communities, and ensures food security for the country in an increasingly insecure international trading environment.

Community-centrism reveals the ways that farming provides significant social values. From the perspective of several farmers interviewed, a focus at the farm level distracts attention from the negative effects that other industries have had on the water system. They used examples, as discussed in Chapter 3, including coal and gas extraction, which have tremendous impacts on the quantity and quality of water. Farmer Tony Piggins believed that farmers were held to a higher accountability than others because they have demonstrated a high level of responsibility regarding environmental care. Piggins warned that coal-seam gas extraction in his area substantially affected groundwater quality, which could result in significant damage lasting thousands of years (even though the company has only been operating there for twenty years).[22] According to farmers, water used in extractive industries does not appear to receive the same kind of attention as farming, even though these industries offer virtually no real benefits to the local communities.

Like extractive industries, the demands of international markets have contributed to a shift in attention away from the value of farming and local communities (and local economies and social structures). There is increasing pressure to produce more for international markets, and secondary industries have begun to disappear. No matter what a farmer produces, they will have to produce much more to secure the same income. Farmer Allen Clark remarked: "If a farm twenty or thirty years ago could grow 1,000 tonnes of wheat, or milk 100 cows or raise 500 lambs and make a living, now they need to be growing 4,000 or 5,000 tonnes of wheat or milking 300 cows or raising 3,000 sheep to support the family on the family farm." In addition, as the price of water increases, so does the price of hay and grain. The price of grain also depends on volatile world market grain prices. Local markets are more stable than international markets, so dairy farmers, for example, try to buy their hay from local farmers.[23] Working with their local communities allows farmers to retain greater value within their communities. Community-centrism focuses on community-based relationships and outcomes to reveal the long-term gains of localised trading structures. Local trade can provide significant economic gains over a longer period and for a wider group of people within the community.

Economic rationalism foregrounds the desire to bolster foreign revenues from the agricultural sector to strengthen the national economy. But a cost–benefit analysis of interventions does not necessarily consider local social impacts or how social impacts might affect economic outcomes over the long term. For example, governments (and farmers) generally agree that the market price reflects the value of the input. Therefore, if the value of the water is

---

22 T. Piggins, personal communication, 2016.
23 A. Clark, personal communication, 2016.

$100, and the government buys it for $100, then all the benefits and costs are built into the value of the water. At the global level, that might be an accurate valuation of the water. At the regional level, it might also be a correct valuation because that reflects the value of using the water. But as soon as the water is sold, it stops being an investment in the community.[24] When there is less water in the community, farmers require fewer inputs, meaning there is less need for services. Further, as the need for government services is reduced, fewer children are attending school in the communities. Therefore, several other values would be gained if the $100 stayed within the community. At a broader economic level, many possible values have not been understood and factored into the analysis.

Russell James of the Murray–Darling Basin Authority explained that there is an interesting philosophical debate in Australia around taking the "free trade, maximise production route". The consequence of this approach might be that, over time, Australian farmers could end up growing a lot of raw feed exports. He acknowledged that such an approach might simply support animal production in Indonesia or China, offering little added value to the Australian economy; however, he believed that politicians were not just interested in the number of dollars each industry generated but also in the multiplier effect of that production system and how many jobs the industries were creating in Australia. As such, governments have started looking at what reduced water availability means for the downstream processing sector. James noted that a more "holistic look" would examine the secondary impacts of reductions in productivity in specific sectors.[25] From these comments, it appears that some government representatives are acknowledging the complexity of economic decisions in terms of impacts on communities both within the farming community and beyond. As this example demonstrates, there is an increasing recognition that community impacts represent an important consideration in policy and planning. The focus on fostering international markets since the 1980s has created significant tensions and has the potential to greatly undermine food security over the long

---

24 G. Knagge, personal communication, 2016.
25 R. James, personal communication, 2016.

run. Community-centrism, therefore, represents a powerful discourse of resistance to economic rationalism.

Community-centrism highlights that successful water management often depends on ensuring strong social networks are in place to determine the effectiveness of water delivery on farms and in areas designated for environmental water delivery. This framing differs from economic rationalism because it sees achieving efficiency as a community endeavour. Community-centrism brings low-cost alternatives to supporting farmers and provides the necessary collaborative frameworks to make collective decisions that benefit the broader community over the long run. Focusing on the community benefits of on-farm efficiency could also reveal the benefits that extend well beyond economics. Farmer Richard Sagwood explained how on-farm water efficiency programs are of value to the community: "saving more water on farms means there is more water to give back to the environment, at a lower cost, and that is a benefit to the larger community and the environment, and to all the taxpayers".[26] Monitoring water delivery is seen as a community-wide responsibility. This broader view does not see on-farm efficiency as simply a gain for farmers but a significant gain for everyone. Strong social networks tend to present increased opportunities for mutually advantageous initiatives like the ones presented here.

A community-centrism discourse assumes the value of long-term government involvement and commitment to communities. From this point of view, withdrawing support from farms and the surrounding communities is problematic. First, as was discussed in Chapter 3, if a third of the water is taken out of the system, the farmers who are left pay a lot more for the water because the value of the water has increased. The second impact is at the regional community level: if the government buys a third of the water, agricultural production is going to be reduced (even when efficiency measures are put in place), less fertilisers used, less grain produced, less trucking required and so on.[27] Farmer Geoff McLeod remarked: "There is a knockdown effect at the regional community level. If there are less farms, there are less

---

26   R. Sagwood, personal communication, 2016.
27   G. McLeod, personal communication, 2016.

children, which means there are less children going to school, which means there are less teachers and less schools."[28] The impacts of an approach that looked at simple buybacks as the primary method of recovering water has had devastating impacts on the rural communities that depend on farming. Farmer Tony Piggins said he had "no problem with governments acting on the need to build sustainability into these farming systems". He believed that sustainability required the government to manage the irrigation process and how water was supplied. He also believed there was a strong disconnect between government bureaucrats' actions and effects at the ground level where the farmer was trying to survive. Piggins and other farmers in his area spent some time consulting with the Commonwealth and state levels of government as part of the water-sharing plan. The farmers warned the government about what would happen if they made severe cuts. Most of the properties along the river where Piggins lives had invested millions of private monies in developing the complex irrigation systems under authorisation and encouragement from the government. He did not understand how the government could simply ask the farmers to walk away from all these investments: "just turning around saying well that's just bad luck, see you later".[29] Farmers like Piggins see the history of government investment as an investment in their communities. They see the government's withdrawal as a betrayal in terms of their long-standing commitment to developing those communities. Community-centrism discourse sees government investment in infrastructure as one element of a longer-term relationship and commitment on the part of governments to the community, and vice versa.

Community-centrism focuses on meeting the needs of the broader community, not just individuals. The focus on individual monetary benefits in the Murray–Darling Basin drove wedges in the community and was seen by several farmers as a poor political process that undermined collective interests. For example, farmer Louise Burge recalled that several of the irrigation companies supported the plan because they thought they would not be able to change government policy, and the money was attractive to them. Similarly, many farmers

---

28   G. McLeod, personal communication, 2016.
29   T. Piggins, personal communication, 2016.

participated in the plan because the short-term gains were attractive, and the drought meant they were in greater need of cash. But, when the government takes half the water out of the irrigation channels through buybacks, all the channels are still in place. All the systems still needed to be maintained, but with half the number of users. Burge explained the problem by making an analogy to the railway transit system: "Think of a railway network, you have got to run the rail system and provide the service for the commuters, but you take x number of paying passengers out, what happens to the people who are left? How much would they have to pay to keep the same system in place."[30] Dismantling nearly half of the region's water infrastructure had devastating economic and social consequences for those who remained in farming. Community-centrism focuses on the overall impacts of taking water out of the broader farming system rather than on individual consequences. Community-centrism also emphasises the value of having farmers discuss, among themselves, how one farmer's actions might affect their neighbours. A focus on individual outcomes had a divisive effect on farming communities as farmers did not have enough opportunities to work together to find solutions that would provide mutually advantageous outcomes. Farmers generally hope that governments recognise the value of farming communities and make efforts to sustain these communities with less water in the system.

In terms of problem definition, the community-centrism discourse posits the value of looking at policies in terms of the way that the broader community is affected. It also uncovers how preserving rural farming communities affects Australian society positively by preserving a way of life and providing a reliable and healthy food supply. In contrast, economic rationalism assumes the market is the most "efficient" distributor of water through "high-value" users. As we have seen, farming has numerous additional values that are not defined and thus calculated within the discourse of economic rationalism. The social values of farming, as opposed to extractive industries, are emphasised by community-centrism. Further, the discourse challenges the logic of promoting exports and offshoring processing and secondary industries. The discourse reminds us of the benefits of

---

30   L. Burge, personal communication, 2016.

keeping businesses in the Murray–Darling Basin. The social benefits of strong agrarian communities include a reliable food source for the broader society; the development of secondary processing and related industries that provide employment and generate revenue; vibrant and healthy rural communities that attract visitors and draw attention to the importance of maintaining environmental sites; and the preservation of the rural culture that is a part of Australia's national identity.

In Chapter 4, I identified the adverse productive effects of what I have termed green environmentalism, thereby helping to articulate an alternative discourse that considers the essential connections between social and ecological systems. Through interviews with participants, I found that farmers engage in a discourse that tries to break through the perceived dichotomy between human beings and nature. Human communities and their welfare are at the heart of effective environmental management. The farmer perspective is more holistic in that it challenges policies informed by the green discourse's over-reliance on human–nature dichotomies. Farmers' conceptions of nature are fundamentally different from that of scientists and government experts. Farmers understand that people and human communities are an essential part of nature. Community-centrism articulates a way of understanding water management that incorporates scientific management and farmer knowledge while recognising the reciprocal relationship between human culture and ecological systems. The discourse recognises that the current situation in the Murray–Darling Basin does not demand a withdrawal of human activity, but a recognition of the critical role people play in managing these ecosystems, particularly the people who live on the land.

Community-centrism suggests that farmers are key environmental stewards, leading community-driven environmental initiatives. Environmental plans depend on the social support of the larger communities of which farmers are a part, but farmers do not always enjoy this type of support. John Bradford believed that people in the cities, for instance, are misinformed about the practices of farmers. He remarked: "people in the city who like to be clean and green, organic and so forth, were making assumptions about what we did and how we used our water. They thought we were environmental vandals." Farmers are disappointed in the lack of recognition they receive for their attention to environmental stewardship. They see themselves within the landscape as environmental managers who help maintain productive and sustainable environments (for farms and wildlife). Bradford points to the on-farm initiatives of fellow farmers John and Shelley Scoullar as an example. They have grown plantations along the creek where numerous birds have found a habitat. But environmental initiatives offer no financial return, and when "you are pressured by not making money and living very close to going bankrupt, you will not go down the path of doing things like that".[31] Community-centrism recognises that the broader community has a role to play in helping farms transition to better environmental practices. The wider society benefits from environmental reforms, but there is a failure among the larger community to recognise the significant role farmers can play as environmental stewards. In contrast, green environmentalism often treats farmers as inherently injurious to natural environments.

From a purely economic perspective, environmental stewardship also depends on the support of the broader community. Community-centrism recognises that community-based environmental initiatives require material prosperity to support them. Where farmers are concerned with their day-to-day survival, environmental concerns invariably take a back seat. Gary Knagge believed that if governments wanted farmers to be more environmentally friendly, then their farms need to be more profitable:

> wealthy people can afford to give back, and we don't have to push the land as hard to make a profit. If we have that extra paddock, then we can afford to be more socially equitable because we don't have our backs up against the wall.

He believed that it was the economic conditions that force farmers to make poor environmental decisions: "when you are struggling hand to mouth something has to give and it will either be you or the land". He added that, while there may be some unscrupulous farmers who do the wrong thing, the more affluent farmers are, the more generous and socially conscious they become. He did not believe it was a question of

---

31   J. Bradford, personal communication, 2016.

whether farmers wanted to be environmentally conscious but whether they could afford to be.[32]

Community-centrism firmly places the intimate relationships farmers have with their land at the centre of how environmental initiatives should be conceived and carried out. Such an approach is quite different from what has happened in recent years in the Murray–Darling Basin. Farmer Shelley Scoullar explained that one of the major problems with the efficiency programs was that many trees needed to be cleared to make room for centre-pivot irrigation. The programs gave preference to the big centre-pivot sprayers over flood irrigation, so they needed to clear trees to get big sprayers into the paddocks. Scoullar recalled that one farmer, whose farm is nearly 20 per cent plantation, was greatly distressed by the need to pull out trees to make room for the centre-pivot sprayers. She recalled him telling her that such actions went against the grain for him because his family had been planting trees and doing everything they could for the environment since they were established.[33] The irrigation group that Scoullar works with was – at the time of our interview – applying for a grant to build nesting boxes for the birds that lost their nests when the trees were cleared. Even though people have planted more trees, they take a long time to get hollow enough to nest in, sometimes as long as 100 years. With this project, they hope to get funding so locals can build nesting boxes to put in the new trees.

Farmers' daily interactions with the land situate them as important environmental land managers. From the perspective of farmers, they routinely engage in environmental management practices: they maintain watersheds, nurture sources of ecosystem renewal, manage local bird and reptile species and maintain ecological processes at multiple levels.[34] The evidence presented here demonstrates that without healthy and prosperous communities, the environment will not be given the attention it deserves. Community-centrism focuses on the community needs that must be met for farmers to shift their focus to their roles as environmental stewards. In this way, farmers can

---

32 G. Knagge, personal communication, 2016.
33 S. Scoullar, personal communication, 2016.
34 Berkes, Colding and Folke 2000.

potentially make a significant contribution to reaching water-related environmental goals.

A central assumption of community-centrism is that solving environmental problems requires that affected parties work together to establish clear and mutually agreeable solutions. In the case of the Murray–Darling Basin, we have seen that there is generally a lack of communication and cooperation regarding environmental initiatives. As we saw in Chapter 4, for example, the government sent scientists to investigate the needs and conditions of bird species in the Murray–Darling Basin. Farmers did not trust the research of the scientists, and they did not feel they had a space to share their knowledge about bird species. Louise Burge, for example, explained that while rice takes up only a tiny space in the farm, it is an excellent crop to grow because of its high biodiversity value, particularly for birds. There are also an estimated 40 billion frogs a year living in the southern rice fields of the basin.[35] Consequently, there is an enormous amount and variety of bird life. The area has many species of migratory birds coming to feed and nest, particularly those that nest in the bush and then come to feed on the frogs in the rice fields. They also have sea eagles, brolgas and many other birds. Farmers have questioned whether hired scientists have adequate local knowledge of these bird populations to go into a community and tell the community what to do. Such an approach arguably entails a top-down process wherein decisions are made without understanding how the area's ecology works and what is realistic. An alternative approach could be for governments and environmentalists to enter a sub-catchment area or valley and ask the farmers there to help them design a solution to protect the species. According to Burge, such an approach "automatically takes the conflict out" because it fosters partnerships so that, depending on the species needing protection, appropriate government funding can be more readily secured.[36] This approach would also give the farmers a space to share information and allow governments to share information with farmers. These outdoor spaces could be sites of knowledge exchange, creating strong social networks

---

35 L. Burge, personal communication, 2016; Myers 2002.
36 L. Burge, personal communication, 2016.

between scientists and the farmers who live on the land and opportunities to cooperate in counting species or running other studies.

When people become willing participants, they are more open to educational opportunities informing them about species. Farmers are also more likely to become engaged in activities like building nesting boxes, monitoring species numbers and other efforts that the government or environmentalists may not have considered. Farmers want to see environmental policy in Australia work from the ground up, and they can play a critical role in determining which studies and projects would provide the most value in environmental initiatives. For instance, creating a community around bird enthusiasts will foster greater care and attention to environmental initiatives. Currently, evidence from my research suggests that governments tend to focus too much on big-ticket items and promise money to large projects that are not well supported by science or evidence. Burge thinks that working with communities is the most cost-effective and intelligent approach.[37] Community-centrism reveals how greater involvement from local communities delivers enhanced environmental outcomes for less money. Further, community-centrism highlights that when communities are invested in outcomes, environmental solutions tend to focus on longer-term outcomes. Evidence suggests that, to date, most of the environmental policies in Australia have been advocacy-based – often made by people at the various levels of government and by environmental groups but with neither group having formed deep connections with the people on the ground. Community-centrism points to this as a failure in terms of long-term environmental sustainability.

Community-centrism reveals that farmers want to help foster strong and ecologically sustainable ecosystems and communities. The discourse shows that even in cases where farmers may not show an interest in environmental outcomes, community-based approaches provide opportunities to engage and nurture an interest in the environment. Giving social capital to farmers will create more opportunities for ecological initiatives and long-term, community-

---

37   L. Burge, personal communication, 2016.

based solutions. The discourse critiques the purely economic rationalist discourse, which assumes that farmers' interests are always individualistic and economically based. Community-centrism draws attention to how financial success and the long-term viability of farms depend on healthy community-based networks. The discourse provides a way of thinking and acting that will help build the kinds of social exchanges that empower farmers and help facilitate more effective environmental management.

Policy solutions

During the Millennium Drought, policy analysts faced the challenges of navigating the complexities of integrating a wide range of alternative views into policies. Overall, it appears that the government has moved towards more polycentric forms of governance in the Murray–Darling Basin. Local entities like councils, regional groups like catchment authorities and governing entities have aligned decision-making with national policy.[38] Governance has depended on a broad consensus about rules, policies and values because no one group was in charge.[39] Further, the knowledge and participation of farmers provided some important answers in developing appropriate policy solutions at this critical time. But my research indicates that many opportunities to gather and apply the local knowledge of farmers were missed. The social relationships among the key players were identified by farmers as a central factor in the effective sharing of information and the development of trust, fostering more successful water-management approaches. Nonetheless, there is a lack of social trust between governmental representatives and farmers in the Murray–Darling Basin.

Polycentric governance structures are affected by the power dynamics at play among competing actors. In the case of the Murray–Darling Basin, hierarchal decision-making structures in which the government privileged the knowledge of some actors above others

---

38   Wyborn, van Kerkhoff et al. 2023.
39   Abel, Wise et al. 2016.

contributed to this lack of social trust. Community-centrism draws attention to the ways that strong social networks allow knowledge transfer to occur. The discourse focuses attention on engagement throughout all stages of the policy development process. For those who adhere to this discourse, problems are defined with community outcomes in mind, and policy solutions are based on these alternative problem definitions. The government can have checklists and measurements for accessing the level of community engagement, but if the government leaves a community without establishing bonds based on a firm sense of trust, the policy, no matter how good it may appear on paper, will likely be met with hostility from the communities that are affected.

Other research on polycentric governance in water management supports the conclusion that it is critical to examine how structures of power affect the capacity to move towards more collaborative approaches.[40] Currently, the trust between farmers and government officials is virtually non-existent. According to farmer Margaret Knagge, when asked what the government has done to help her and her husband through the drought, she replied:

> What has the government done to help? Nothing, in capital letters, but actually worse than nothing. They have spent a fortune and are in the process of spending a fortune to achieve nothing, to hinder the progress of farmers, to obstruct the work of farmers by using a paradigm or a worldview which is based on extension research dating back to the 1960s and 70s when we were in a world of productivity increases.[41]

In community-centrism, re-establishing social trust is the key to recognising long-term solutions to the problems facing the Murray–Darling Basin. If community-centrism was widely accepted or brought into dialogue with the other perspectives in deliberative processes, the power dynamics that negatively affect efforts towards

---

40 Colloff, Grafton and Williams 2021; Da Silveira and Richards 2013; Gaventa 1982.
41 M. Knagge, personal communication, 2016.

polycentric governance could be overcome. Alternative policy choices may arise through this alternative discourse. For example, farmers could work together to implement environmental initiatives to manage water more effectively through inter-farm agreements. Farmers could identify opportunities to work together to develop solutions they could put forward to governments, thereby reducing the need for governmental interventions. In addition, governments have also begun to identify ways to better engage in communities, generate collaboration and utilise more effective policy instruments.

When farmers themselves come together to build social relationships, the effects can be profound. Strong social networks can lead to increased knowledge production and transfer. Farmer John Hand told me that irrigators are improving the environmental corridors around their properties. For instance, a corridor runs through his property along a creek that was once used for flood irrigation. The farmers have been steadily planting trees in this area to develop another environmental corridor. They have also tried to tie several vegetation areas together so that the corridors link up. As the trees mature and the corridors are connected, there are increasing numbers of bird species in these areas. He recalled he had recently seen at least ten new varieties of honeyeaters that he had never seen before. He has also seen the endangered bittern return to the area.[42] If farmers worked with governments and environmental groups to ensure these corridors extended to national parks and other private residences, the environmental effects would be even greater. They could also work together to monitor the health of these corridors both on and off the farms.

Community-centrism is about leveraging social relationships. Farmers demonstrate this approach through various initiatives and provide examples of how this can be done effectively. Water reform has meant that farmers must find ways to do more with less water but, in some cases, they have harnessed the power of their social networks to manage through difficult periods. Shelley Scoullar, John Hand and John Bradford decided to pool their water resources for one year to grow a small rice crop, as opposed to not being able to grow any rice

---

42   J. Hand, personal communication, 2016.

at all. In previous years they had combined water but not to the same extent. They were determined to grow rice because the crop provides many benefits for their other crops. When they put a cereal grain on top of their fallow rice ground, it gives the cereal a good start. The rice allows them to maintain a good crop rotation so that they get two crops from the same water.[43] The social relations between these farmers allowed them to remain viable and survive through the tough years. Community-centrism reveals that leveraging social relations can give farmers more opportunities to survive drought.

Governments can also learn how to capitalise on social relationships to produce better outcomes. Governments, for instance, could intervene by bringing farmers together to form their own solutions and pool their resources to remain viable through the dry periods. In the past, instead of calling all the farmers together to discuss the plans of the Murray–Darling Basin Authority, the Commonwealth government could have begun by first bringing the farmers together to explore possible solutions among themselves. Such discussions could be facilitated and even guided by the government but would be more effective because they would give farmers a chance to find their own solutions, and they would have ownership and accountability over those solutions. Government officials would also be able to explore different options they had not considered earlier. These would be just some of the potential productive effects of taking seriously the priorities emphasised by the community-centrism discourse.

Table 5.1 summarises some ways that the problem definitions of government representatives and farmers diverge. Further, it reminds us of the policy options and alternatives that arose from the point of view of the respective parties. The community-centrism discourse summarises many of the views of farmers, while the views of government representatives are articulated throughout the previous chapters.

---

43   S. Scoullar, personal communication, 2016.

Table 5.1 Government and farmer problem definitions and policy solutions

Problem	Government problem definition	Governments' policy solution	Farmer problem definition	Farmers' proposed policy solution
Salinity	Not enough water Over-allocation	Divert water through administrative and legislative means Release more environmental water at the right times	Invasive fish species Overdevelopment along riverbanks Overwatering	Programs to improve water quality, i.e. reducing invasive fish populations, revegetation programs
Drought	Not enough water Over-allocation Climate change	Divert water through administrative and legislative means Buybacks Dictate water diversions Enforce market-based penalties	Irregular weather patterns as part of Australian conditions	Identifying locally specific water needs Community-led initiatives to conserve water Long-term drought mitigation projects
Flooding	Climate change Administrative errors Poor infrastructure	Farmer investment in infrastructure Modernisation projects	Government releasing too much environmental water from the dams	Publicly funded infrastructure Government accountability framework for third-party impacts of increasing flows
Over-allocation	Water as the property of the Crown and highest buyers Water as a commodity or	Buybacks Market-based instruments that deliver the highest monetary value for water Increasing regulations	Water is the property of farmers and communities Long-term values of water	Securing existing water allocations Reducing new allocations Securing allocations in established communities

Problem	Government problem definition	Governments' policy solution	Farmer problem definition	Farmers' proposed policy solution
	financial instrument			
Environmental damage on farms	Farm-level problem	Farmer responsibility	Community-wide problem	Co-led farmer–government initiatives to preserve ecologically sensitive areas on farms
Environmental damage to wetlands	Government-level problem	Increasing flows to wetlands	Community-wide problem	Rehabilitation projects
Farm debt	Farms are inefficient	Increase production through greater efficiency	Community-wide problem Effects of secondary industries on the Australian rural economy	Interest payment relief Deferral of interest payments Loan repayment deferral Long-term restructuring plans
Economic hardship in rural communities	Rural community problem	Rural community responsibility	Shared responsibility of governments and farmers	Government-funded community support services Long-term economic restructuring

Much of what I have discussed in previous chapters reflects a top-down approach on the part of the government. I also uncovered evidence that some in government hope to see a change in this approach. They want governments to start to look towards communities to ask how change can occur. Such a reorientation could allow for a more deliberative and bottom-up approach to policy development in the basin. For example, Tony McLeod (General Manager, Water Management, Murray–Darling Basin Authority) told me that he believed that the Australian system of sharing water between people within a highly variable system, while having relatively low amounts of regulatory supply, is among the best in the world. He believed that the systems in place of rationing and sharing water are incredibly well developed and that, in the decade since 2006, they have moved towards a more sustainable footing through the basin plan. He also told me that the Australian "water-user community" understands all the nuances of water and that "people are highly articulate on these matters". McLeod added: "the collective wisdom of the water users is not to be underestimated either; in fact, it's vast, they literally bet their farms on water access". When speaking with the irrigation community, McLeod is always impressed with the knowledge the community has about water management. He believes that "the leadership of the sector realises the benefits of understanding all these moving parts", and that they "also realise that you cannot govern how much it rains".[44] These statements demonstrate a recognition among some in the government that the farm community relies strongly on the collective wisdom of farmers, and they want that wisdom to be better utilised. It would appear that, at least for some in government, farmers have made an impression. McLeod's comments highlight the significance of social trust in forming the kind of open dialogue needed for deliberative democratic engagement. His voice appears to represent a divergent view from many in government as he tried to articulate a recognition of farmers' contributions as an important step towards open dialogue and deeper understanding.

Social networks can be fostered between green advocates, government officials and farmers. Folke and colleagues have

---

44   T. McLeod, personal communication, 2016.

written that social memory and social capital should be strengthened for community-based conservation to be successfully incorporated.[45] The discourse of community-centrism fosters engagement in communities and helps improve trust among parties. The policy enforcement mechanisms that develop as a result of the discourse tend to reinforce trust, encourage a sense of community and support social networks among the various stakeholders.[46] The environmental values of the larger society can thus often be seen reflected at the farm level. But it must be understood that these types of initiatives depend on the support of the broader community. If society values efficiency and low-cost production above all else, these values influence how farmers treat their land. If productivity is the primary value, there is little incentive for farmers to increase their environmental efforts. Further, farmers require social feedback for their efforts. Successful environmental management requires that farmers, government representatives and environmental advocates join together and build social communities to effectively communicate their efforts and ideas with one another. Table 5.2 summarises the characteristics of community-centrist discourse in the Murray–Darling Basin.

Conclusion

As is discussed in this chapter, community-centrism draws attention to the value of retaining strong rural communities. Farmers do not act alone; they depend on local towns, a reliable labour force, multigenerational continuity and strong familial or personal support structures to have the support and desire to remain in farming. The interviews with farmers demonstrated that community-based outcomes were a key motivating factor in their lives. This sense of community and connection can be extended to how farmers view the land on which they live. The farmer is ideally situated to observe all the holistic connections that exist over an extended period and in complex natural environments. Farm-level outcomes depend on a wide range of

---

45   Folke, Hahn et al. 2005.
46   Ruiz-Mallén and Corbera 2013.

Table 5.2 Community-centrism: transcripts and policy, legislation and actions in the Murray–Darling Basin

Transcripts, metaphors and rhetorical devices	Policy, legislation and actions taken
community bottom-up approach collective wisdom environmental stewards grassroots local knowledge knockdown effects community-based solutions localism	Various community-based water level monitoring and reporting Strategic community-determined water buybacks Nesting boxes initiative Farm-based initiatives to control invasive species Strategic research that engages local farmers and communities throughout the process

local variables, such as soil type and water-table levels. In addition, the value of knowledge is determined by the social relationships of actors, and these relationships must be strong. The relationship between the farmer and government expert, for example, can determine how much governmental policies will influence a farmer's decision-making. If the farmer does not believe that the government representative is fully invested in policy outcomes, the farmer has little reason to trust government information or recommendations. Further, the farmer sees it as the government's responsibility to work towards gaining that trust through direct involvement in the community. The farmer, therefore, must believe that the actors providing information are heavily invested in the potential outcomes of acting on that advice. Although government gives more weight to scientific knowledge in policymaking, this knowledge is only helpful if it can be applied to local conditions. More than an initial investment in science and technology is needed; understanding the effectiveness of a policy measure from the farmer's perspective is essential. Community-centrism brings locally based community knowledge to the fore.

By focusing on the social life of rural communities, community-centrism helps us recognise how social outcomes have significant economic and environmental consequences. Farmers' daily interactions with the land situate them as important environmental land managers. Farmers in the Murray–Darling Basin routinely employ various ecological management practices.[47] But none of this would be possible without the support of the broader community and the resources to carry out these activities. Culture, the sets of practices and beliefs that are developed through human communities, determines the ways that we interact with the land. Culture provides a set of instructions about how to live on the land. The way we treat the land is representative of the health of human cultures, *and* human social relationships are a significant determinant of environmental outcomes. Murray Bookchin argued that the roots of ecological problems are closely tied to human social problems and can be solved by reorganising society along more ethical lines. Bookchin's approach acknowledged the co-dependent relationships between human societies and natural systems.[48] Social justice is at the very heart of environmental justice. Without a community-based approach, farmers *and* the environment are likely to suffer.

As the examples in this chapter demonstrate, the potential contributions of farmers in terms of a practical and integrated approach to water management practices are not being fully realised. By analysing community-centrism and its potential to shape bureaucratic planning, we can shift the conversation about water reform towards the interactions between social, ecological and economic needs.

---

47 Berkes, Colding and Folke 2000.
48 Bookchin 1994.

6
Policy alternatives

This work addresses how environmental discourses shape the parameters of acceptable policy choices in the Murray–Darling Basin and subsequent outcomes. I asked what the defining discourses were in the Murray–Darling Basin and how some discourses have gained more influence than others. I asked what forms of knowledge these discourses legitimise and how this affects policy. I inquired as to what alternative perspectives, knowledges and policy options are excluded, and what would be the policy implications of these alternative perspectives. As we have seen, administrative rationalism has significantly affected government policymaking in the Murray–Darling Basin. At the same time, the impact of economic rationalism in Australia, closely associated with the era of neoliberalism in much of the rest of the world, has set the parameters for acceptable social interventions in crises. Governments try to navigate the interests of farmers by putting in place democratic processes and institutions meant to facilitate a more engaging and responsive approach. But these processes and institutions often overlook key obstacles to meaningful engagement and ignore important issues and barriers to understanding. One such barrier is conceptions of the environment, which inform how we think and address problems. Understandings of nature as separate from people have had a profound impact on the way policymakers develop environmental interventions to deal with crises. Throughout these pages, it has been argued that there is a need to understand how different discourses shape *and* limit our capacity to respond to environmental problems. The troubles in the Murray–Darling Basin call for an approach that incorporates the knowledge of farmers and the science-based approach of the Murray–Darling Basin Authority. A deeper understanding of how discourses shape decision-making is needed. Each adaptation option and any decision about the governance and institutional arrangements for adaptation is underpinned by values associated with ideas that determine what are considered worthwhile adaptation actions and what are not.[1]

A key challenge in the region will be to develop positive collaboration between the government and farmers so that the interests of all parties can be addressed in the long term. But since the Millennium Drought, trust has been dramatically undermined. Confidence could be restored by recognising the capacity of farmers to play an integral part in environmental management. While managing water effectively in the current circumstances is challenging, an approach that highlights social values and community needs would significantly redefine regional policy development.

In sum, we need to focus on the alternative discourse that has developed in the farm community. Social ecology, the critical social theory founded by author and activist Murray Bookchin in the 1960s, provided the foundations for helping me identify another way of understanding these issues. Social ecologists critique social, political and environmental trends, advocating for the reformation of society along ethical, ecological and community lines. Building on social ecology, I have presented a community-centrist discourse that provides an alternative way of understanding social and environmental issues and encourages more direct democratic engagement.

Early on, I discussed the importance of administrative rationalism in shaping how governments have approached policy in the Murray–Darling Basin. While administrative rationalism is often defined by a tendency to rely on scientific evidence and expert

---

1   Leonard, Parsons, et al. 2013; Spence, Poortinga et al. 2011; Wolf and Moser 2011.

knowledge, much of the information gathered and relied upon in administratively "rational" processes can be problematic for being either incomplete or based on evidence from limited sources gathered over inadequate timeframes. It appears that the Commonwealth government tended to rely more substantively on evidence that supported the plan instead of seeking out the best and most diverse sources of knowledge. There was a strong political imperative to support the plan. Consequently, perhaps without an entirely conscious motive, government actors sought out science that supported the plan. After speaking about the process with farmers, civil servants and some scientists, it is clear that the process was far too rushed to have considered the full scope of scientific evidence. As Colloff, Grafton and Williams argued, there is a strong inclination towards "administrative capture" in the case of the Murray–Darling Basin.[2] The authors found that publications and public comments were controlled by decision-makers' contracts, intellectual property rights assignments, and control over what information was allowed to be publicly available, with no time for extended analysis and certainly not with strong feedback loops. But no one among my interviewees disagreed with the necessity to keep the rivers healthy. Farmers characterised the government's reaction as "knee-jerk" or "reactive" instead of deliberative and scientifically grounded. Further, when I spoke to government representatives, there was a tendency to prioritise scientific over locally based knowledge. Discussions with farmers reveal that, while playing a vital role in policy development, science cannot be used in isolation from other forms of knowledge, particularly farmers' lived experience and local knowledge.

The dangers of relying too much on modernist and centralised planning are not limited to communist states; democratic governments and the bureaucratic institutions within them that develop an autonomous identity are also susceptible to pushing their own agenda, which can undermine democratic values and decision-making. Arturo Escobar has commented in some detail about the dangers of high modernism in democratic states:

---

2   Colloff, Grafton and Williams 2021.

Perhaps no other concept has been so insidious, no other idea gone so unchallenged, as modern planning. The narratives of planning and management always presented as "rational" and "objective" are essential to developers. A blindness to the role of planning in the normalisation and control of the social world is present also in environmental managerialism.[3]

Some would argue that administrative rationalism has largely been supplanted by economic rationalism as the dominant discourse. My research suggests that in the case of the Murray–Darling Basin, both have a strong influence over policymaking, and in many cases, one serves to reinforce the other. We saw from many examples that economic rationalism created a need for administrative rationalism where it may not have existed in the past. For instance, buyback programs and the commodification of water as a tradeable asset created the need for more government oversight.

Interestingly, economic rationalism was a dominant discourse among farmers and governments alike in rural Australia. In both cases, economic rationalism limited how farmers and government representatives understood the problems in the Murray–Darling Basin and the kinds of solutions they envisioned. Farmers internalised the discourse of economic rationalism and, for the most part, believed that the financial hardships they encountered were primarily the result of their own shortcomings and not the economic reality and the resource limitations they encountered. It was easy to point to the few farmers who remained afloat during the Millennium Drought and think that the rest must have done something wrong. Further, even though farmers generally shared the same ideas as governments regarding their internalisation of economic rationalism as an ideological framework, government officials often characterised farmers differently. Some officials accused farmers of "simply looking for a handout". Because farmers internalised this rhetoric, they avoided any type of policy reform that remotely resembled government assistance, even in cases where such assistance could result in government profits in the longer term.

---

3   Escobar 1996, 50.

My research focused on two communities, one around Griffith and the other around Finley. In Griffith, the Murray–Darling Basin Plan was met with deep hostility; in Finley, a mixed reaction took place, with some farmers benefiting from the Millennium Drought. Generally, the responses were close to what I imagined they would be. The level of distrust and anxiety about government interventions was very high; farmers tend to be opposed to government support. At the same time, the government tends to portray farmers as heavily reliant on support. One government official labelled the farmers "whingers". Another government official said that successful farmers do not complain but that there is a "welfare mentality" among some farmers. From my observations, this signified a dismissive attitude and represented a false characterisation. There is a popular disdain for welfare among the public in Australia, and it has become a point of contention which governments have successfully rallied people around.

What often gets lost in these discussions is the very real and sometimes tragic circumstances that farmers face. The realities of international competition, low food prices and high input costs make farming an unstable and unpredictable business. We can lose sight of the vital need farmers fill; without them, we would simply not eat. For this reason, offering some special considerations and incentives makes sense. There are social values associated with farming, and the responsibility falls on all of us to ensure that agriculture is maintained as an economically viable enterprise. In the end, we all suffer under the weight of an industrial food system that does not consider human and environmental values. As Bill McKibben wrote:

> If the damage to community is arguable, an industrialised food system has other costs that are both more prosaic and more obvious. Part of the reason for that low, low price for food is that we pay many fewer farmers a smaller percentage of our food dollars.[4]

In a world of increasing scarcity, malnutrition and drought, we have a shared responsibility and incentive to ensure the success of farmers.

---

4   McKibben 2007, 54.

Economic rationalism puts the success of farmers squarely on the individual farmer and leaves entire communities that rely on those farmers vulnerable to economic ruin. If we are to survive and flourish as a society, we must understand that our collective success depends on the success of individual farmers. Their failures will invariably affect every one of us.

While individualism is a key aspect of economic rationalism, individualism also influences democratic pragmatism. In Western societies, democracy is understood to mean that everyone makes the best decision for themselves, and the majority wins. This approach has a particular rationale that gives the individual a special position. But this focus on individualism can have the effect of diminishing efforts at collective decision-making. The emphasis on individual rights and needs detracts from the focus on collective outcomes. Negotiations have become a zero–sum game where we believe that winners and losers are an inevitable outcome of the decision-making process. But within the confines of the capitalist political economy, the scope of democratic authenticity is limited. As was discussed in detail in the section on democratic pragmatism, there were important "constraints" to realising a truly authentic process of democratic engagement. In *Deliberative Democracy and Beyond,* Dryzek introduced the idea of a more authentic deliberative democracy based on a particular set of principles that attempt to expand the scope and depth of deliberation within democratic states.[5] Inspired by Dryzek's recommendations, findings from the literature and farmer interviews, I have identified some guiding principles of deliberation with farm communities while recognising that local circumstances dictate that it is important to avoid being overly prescriptive.

In the realm of deliberation, it is imperative to acknowledge that effective communication extends beyond the confines of purely rational arguments. Embracing alternative forms of expression, such as rhetoric and storytelling, can serve as gateways to more authentic deliberation. For instance, casual one-on-one meetings between farmers and government representatives can lay the groundwork for improved communication in more formal settings.

---

5   Dryzek 2000, 29, 163–75.

Moreover, it is crucial to dispel the notion that rationality equates to neutrality. Recognising the diversity of discourses is essential, as so-called "rational" discourse often aligns with the most dominant voices. By emphasising the contestations between different discourses, marginalised voices gain a platform, increasing the capacity to challenge hierarchical power relations. There is a necessity to broaden the scope of voices heard during deliberative processes. Even views that may diverge from established democratic norms, such as forms of prejudice or sectarianism, should not be silenced. Faith in the deliberative process involves the belief that dissenting opinions can, through open dialogue, be positively transformed. Unpleasant biases between farmers and governments, for instance, must be confronted openly, as exposing these views is essential for paving a way forward.

Deliberators are often expected to align with certain values to be admitted to the process, but this should not preclude anyone from participating. Deliberation is a skill that requires practice and improvement, and all participants should have the opportunity to cultivate values such as respect and reciprocity. The example of the Murray–Darling Basin underscores the importance of ensuring that participants – in this case, farmers – are adequately prepared and supported in unfamiliar deliberative processes initiated under tense circumstances.

Authentic deliberation allows parties involved to gain awareness of each other's interests and the collective interests of vested groups. Shifting the focus from a zero–sum game perspective to one that prioritises collective interests reframes the deliberative process. This perspective, as exemplified in the discussion on community-centrism, underscores the significance of community-based outcomes that are often overshadowed by dominant discourses.

Challenging the misconception that deliberation should be confined to existing representative institutions, it is also crucial to consider alternative avenues for democratic participation. The section on democratic pragmatism in Chapter 3 elucidates how various disciplines limit authentic democratic participation within states. Acknowledging the weak deliberative position of isolated farmers and empowering them through community support demonstrates that influence and administrative power can extend beyond traditional institutional frameworks. Recognising the inherent inequality among participants in deliberative processes is paramount. Although complete rectification of power dynamics is impossible, emphasis must be placed on the right to participate. Expanding the range of permissible communications becomes instrumental in fostering a more inclusive deliberative process. In the context of the Murray–Darling Basin, diversifying communication tools and platforms beyond town hall meetings is imperative, considering the constraints these limited platforms impose on farmers.

Finally, evaluating the strength of a deliberative process goes beyond its ability to deliver individual policy outcomes. The enduring benefits of a careful and inclusive deliberative process lie in the legitimacy it affords participants, irrespective of the final policy decisions. Building bonds of respect and reciprocity throughout the process creates a foundation for more positive future deliberations. Moreover, the process can unearth the underlying values that motivate actors, paving the way for alternative policy choices that address fundamental values rather than merely immediate concerns.

While I observed a diverse range of opinions about the necessity of the basin plan, most agreed that the process was deeply flawed. There are deep divisions within the farming communities regarding the types of approaches that the government should use. Those most dissatisfied with the plan, particularly farmers who experienced severe damage due to flooding, pointed to mismanagement as the cause of the problem. At the time of the interviews in 2016, farms were being flooded, and farmers were experiencing significant losses. Even though 2016 was the wettest year in more than a hundred years, farmers were unsure about getting a full water entitlement for the spring–summer season. While still firmly opposed to the plan, more moderate farmers were trying to work with the Commonwealth government to get what they needed, as they saw it as impossible to fight the government. Farmers reported that they were trying to find a viable approach to dealing with the government but also trying to achieve results they could live with and send a strong message that communities were being damaged.

From the outset, farmers and governments assumed that there could be no collective wins through the negotiation process. The Commonwealth government knew what it had come to do and what it hoped to achieve before the process even began. Farmers inevitably responded by digging their heels firmly into the ground and hoping to gain the most from the process. Such a reaction, I believe, was an inevitable consequence of the government's posturing from the outset. If it had chosen to take an approach that demonstrated an openness to alternative suggestions and even to potentially change its goals, farmers would have been more receptive to the entire process. From the perspective of farmers, the efforts at democratic decision-making were largely top-down and did not take into consideration views that were outside of the structured consultative agenda. The consequences of such an approach, as explained throughout the section on democratic pragmatism in Chapter 3, were a breakdown of trust between all parties involved and a failure to incorporate valuable insights that fell outside the government's agenda.

There were important "constraints" to realising a truly authentic process of democratic engagement. The research on water governance strongly supports polycentric approaches, particularly more collaborative approaches.[6] However, power dynamics can interfere in efforts towards more collaborative approaches.[7] Focusing our attention on these power dynamics and how to address them in practical ways could contribute significantly to more collaborative approaches. Further, while some scholars examine discursive factors that may limit the capacity for polycentric governance,[8] there have not been attempts in the literature to characterise how a specific group of stakeholders, like farmers, construct their understanding of the issues. This detailed discursive analysis situates the voices of farmers as central to understanding the conditions under which they may be engaged in the collaborative process. The analysis of farmer discourse in this case offers insight into how those affected by policies understand the impacts of these policies on their lives. Resistance to the dominant discursive framings of an issue is better understood by uncovering the assumptions and underlying motivations of individual actors affected

---

6   Ansell and Gash 2008; Garrick, Heikkila and Villamayor-Tomas 2018; Pahl-Wostl, Arthington et al. 2013.
7   Behagel and Arts 2014; Zeitoun and Allan 2008.
8   Wyborn, van Kerkhoff et al. 2023.

by policies. For governments to reconcile competing interests and values, they must understand the underlying motivations of actors.

We observed that while there may be attempts to include a plurality of voices in the deliberation process, not all voices are treated with the same value. Collaborative approaches are grounded in the assumption that all actors can contribute to the process and affect outcomes. Where actors have equal power, this may be possible. But, in the case of water governance, the actors who come together in the collaborative process are rarely equal. Some actors have more power to influence the collaborative process than others. Further, the state remains the dominant decision-making authority.[9] So-called "rational" discourse is often associated with the most powerful voices and actors.[10] Looking at the ways discourses can both silence and empower actors, as is attempted throughout these pages, allows for marginalised voices to be heard and for hierarchal power relations to be challenged. In model deliberative processes, participants are open to changing their opinions through persuasion. Deliberative processes are characterised by respect, sharing of information and allowing all actors to participate freely.[11] This means that everyone should have the opportunity to participate in the process and learn skills like cultivating respect and reciprocity as needed in order to actively stay engaged in the process.

---

9   Brisbois and de Loë 2016.
10   Dryzek 2000, 163–75.
11   Dryzek 2000.

Conclusion

Farmers play a crucial role in environmental stewardship. I propose a bottom-up approach to environmental planning, emphasising the value of local knowledge. The focus is on deploying institutional arrangements and policy instruments to support community-based goals and capacities.[1] Community-centrism represents a departure from much of the current literature on the Murray–Darling Basin. For example, some scholars have argued that Australian taxpayers will be worse off due to the community consultations undertaken by the Murray–Darling Basin Authority.[2] Ross, Buchy and Proctor have argued that consultations were problematic and that risks may have outweighed benefits: "consultation burnout"; the capacity to "raise unreasonable expectations"; the possibility that the most powerful stakeholders shape the issues, thereby limiting input from less powerful stakeholders.[3] Similarly, Garrick, Whitten and Coggan believe that the costs of consultations were too high and that if governments had limited public consultation and pursued their agenda of market acquisition unimpeded, there would have been much lower transaction costs associated with the reform process.[4] Grafton and Horne

---

1   Robinson, Bark et al. 2015.
2   Crase, O'Keefe and Dollery 2014.
3   Ross, Buchy and Proctor 2002, 216.

questioned the value of local knowledge, claiming that farmers do not grasp the extent and complexity of the problem beyond their own farms. They argued that farmers did not have the ability or knowledge to see the larger environmental picture.[5] The authors believed that the government had the tools and resources to manage water properly; it just needed time to perfect the policy instruments it was trying to implement. They wrote: "in Australia, developing a clear and transparent regulatory framework has taken about two decades ... Time, patience, persistence and effective governance arrangements underpin the emergence of robust water markets in Australia."[6] Grafton and Horne have reasoned that, while local expertise can contribute to policies like environmental recovery strategies, local input can undermine the emergence of strong water markets: "independent development of water resources within each state in the southern Murray–Darling Basin led to the over-extraction problem that currently exists".[7] This work has questioned these assumptions, particularly concerning the benefits of high-level bureaucratic management by the Commonwealth government. This research has instead articulated the value of pursuing policy change with a greater commitment to communities.

It is difficult to grasp how the social structures dictating our everyday lives also limit our choices. As an ideological framework, social ecology envisions a moral economy that moves beyond scarcity and hierarchy towards a world that re-harmonises human communities with the natural world while celebrating diversity, creativity and freedom. The theory suggests that the roots of current ecological and social problems can be traced to hierarchical modes of social organisation, and these hierarchies cannot be resisted by individual actions alone. They must be addressed by collective activity grounded in radically democratic ideals. My research aims to cultivate a deeper awareness of how culture moulds our perception of nature and

---

4   Garrick, Whitten and Coggan 2013.
5   Grafton and Horne 2014.
6   Grafton and Horne 2014, 69.
7   Grafton and Horne 2014, 69.

environmental challenges, recognising the limitations such an awareness imposes on potential actions.

The case of the Murray–Darling Basin presents a critically important question about how we should think about environmental management and long-term sustainability. Agriculturally productive areas of rural Australia are generally not considered "natural" environmental habitats by governments or city dwellers. What most people do not understand is that human interventions – like the barrages that were erected along the Lower Lakes to create an artificial freshwater system or the dams further upstream – have led to such massive changes in the landscape that it is impossible to determine with any certainty what constitutes a "natural environment". In his 1989 book, *The End of Nature*, Bill McKibben wrote that it would soon become impossible to consider any part of the natural world as separate from people. The idea that only the environments that are protected from the impacts of people can be conceptualised as "natural" is deeply problematic in a world where every millimetre of space, if not put to productive use, is in some other way affected by human activity. The protected areas of the Coorong and Lower Lakes, the regions the Commonwealth government spent nearly $13 billion to protect, are not natural ecosystems. These environments have been affected by Western development going back over a hundred years. It is necessary to reflect on how our relationship with nature has evolved. It is foolhardy to believe that nature is something outside us that needs to be protected from us. It is only through an acceptance of our worldwide impact that we can form a new basis of understanding that emphasises the potential symbiotic relationships that can exist, as opposed to the opportunistic relationship of exploitation. This change in orientation does not mean that we should not protect certain environments or that we should use nature to our mutual advantage wherever possible, but rather we should understand better the dynamic relationships that exist both in "nature" and in productive environments, *both* of which are touched by human activity.

Conceptualisations of human beings as outside nature not only misguide us but also undermine our role as caretakers of the Earth. As Arturo Escobar so eloquently understood, nature is constructed through the full range of human activities and the cultures and technologies that define those activities:

> Although many people seem to be aware that nature is "socially constructed", many also continue to give a relatively unproblematic rendition of nature. Central to this rendition is the assumption that "nature" exists beyond our constructions. Nature, however, is neither unconstructed nor unconnected. Nature's constructions are affected by history, economics, technology, science, and myths of all kinds as part of the "traffic between nature and culture".[8]

As discussed in Chapter 4, there is a danger of advancing romantic images of natural environments that in no way represent the historic character of a given place. As Bruce Pascoe explained in his book, *Dark Emu*, there is very little about the modern Australian landscape that resembles the natural environment in which Australian Indigenous peoples lived and farmed for thousands of years before the European settlers.[9] There is a real danger in rewriting history and imagining natural spaces in ways that in no way represent the past. As Kate Soper explained:

> Ecological critics of the atomizing and destructive effects of instrumental rationality need to be careful in redeploying the organicist imagery that has been such a mainstay of right-wing rhetoric. Romantic and aestheticizing approaches to nature have as readily lent themselves to the expression of reactionary sentiment as a sustained and radical critique of industrialism, and this means that left-wing ecologists, however understandably keen they may be to re-size this tradition of romanticism for their own purposes, are dealing with a problematic legacy.[10]

---

8   Escobar 1996, 64.
9   Pascoe 2018.
10   Soper 1995, 150.

A key challenge in the Murray–Darling Basin will be to develop positive collaboration between governments, farmers, green advocates and other key stakeholders (including Indigenous peoples). In this way, the interests of all parties can be addressed in the long term. However, since the Millennium Drought, that trust has been dramatically undermined. Trust could be restored by adopting an approach that recognises the capacity of farmers to play an integral part in environmental management. While managing water effectively in the current climate is challenging, an approach highlighting social values and the community's needs could contribute to a significant rethinking of regional policy development. While collaborative approaches to water management try to incorporate as many people as possible within the consultation agenda, community-centrism focuses on the nature of the relationships between these people. It focuses on building social trust and deconstructing hierarchal power structures. As some of the examples in this work demonstrate, power and perceptions of power can interfere with more collaborative approaches to polycentric governance. Our worldviews are often constructed through discourse, but it is impossible to ignore the role of power dynamics in determining which discourses gain currency.

Authentic deliberation among parties allows them to become aware of the interests of other parties and the collective interests of vested groups.[11] Often, negotiations are viewed as a zero–sum game, with the interests of individual actors in competition with one another as the focal point. Instead, centring on collective interests tends to reframe the deliberative process positively by focusing away from the traditional emphasis on individual interests. It is a mistake to believe that the deliberative process should be relegated to the existing representative institutions or the legal systems of democratic states. As is discussed in detail in the section on democratic pragmatism in Chapter 3, there are many disciplines that limit authentic democratic participation within states. Free trade and mobility issues associated with the demands of living in a capitalist state often contribute to these types of disciplines. Civil society and the public sphere are important spaces for deliberation. A community-centrist approach reminds us

---

11   Dryzek 2000.

that farmers are not facing problems alone; not only do they support whole communities, but they have whole communities available to them in their attempts to influence the deliberative process.

In this work, community-centrist discourse is presented as an alternative to the dominant discourses that have characterised decision-making in the Murray–Darling Basin, particularly the green discourse that has been dominant among environmentalists like Tim Flannery and Richard Kingsford, among others. This work has advanced the perspective that farmers can make a significant contribution to environmental stewardship. I have advanced a bottom-up approach to environmental planning and management wherein local knowledge and planning are given greater value. This orientation focuses on how institutional arrangements and policy instruments might be deployed to support community-based goals and capacities.[12] Community-centrism places human social relationships at the heart of environmental decision-making, providing a reconceptualisation of environmental problem-solving and surfacing economic, environmental and social opportunities. Building on the insights of Murray Bookchin, Elinor Ostrom and others – but grounded in the voices of the farmers I interviewed – community-centrism focuses on the role of community-based engagement.

Many of the examples in this work illustrate how hierarchal social structures and decision-making have hindered the capacity of governments to address the crisis in the Murray–Darling Basin in a way that is responsive to both human communities and the environment. Social ecology encourages a transformative position on social and environmental issues and promotes direct democratic action.

The effectiveness of adapting to climate change depends on the cultural fabric of the group involved in the implementation process. An acute awareness of the ways in which culture shapes our understanding of nature and environmental problems and can limit the possibilities for action is needed; this has been the aim of my research. Social practices, beliefs and ideas are at the heart of how we problematise our relationship with nature. Human cultures, and the social interactions between people, are determining factors in the ways that we relate to

---

12   Robinson, Bark et al. 2015.

nature. As Kate Soper explained: "Our developed powers over nature have brought about a situation in which we are today far more at the mercy of what culture enforces than subject to biological dictate."[13] Soper argued that many miseries afflicting the world could easily be eradicated if not for the entrenched social conditioning responsible.

Change is often determined or influenced by those who have social capital. As was discussed in the section on administrative rationalism in Chapter 3, scientists and experts have social capital that governments rely on to gather support from the broader community. But other people can work towards gaining social capital through activism and networking. These actors can have a tremendous influence on the nature of discussions. In this case study, I found many of these strong voices among the women in the farming communities. It was apparent that male farmers often tended to take on all the personal responsibility for the failures they experienced in their businesses, despite external conditions. Their wives tended to get angrier and more vocal about how outside influences had affected them. Many of the women who remained took strong leadership roles. They relied on the social capital they gained from being such a strong and stable part of their local communities. One of my most enthusiastic interviewees, Helen Dalton, became a member of the New South Wales Legislative Assembly and has represented Murray since March 2019. She is a member of the Shooters, Fishers and Farmers Party and has since 2016 been working diligently to gain media and government attention for the Murray–Darling Basin. It is also interesting to note that all the farmers I spoke with wanted to be identified in my research findings. They have never felt heard by the government, the media or even researchers. They clearly understood the need to gain social capital by communicating their needs.

During the Millennium Drought, many people were touched by suicide, divorce, bankruptcy and other problems. While many acknowledged that the government could not stop the drought, they thought more could have been done to maintain the social fabric of the communities. I have argued that while the actions of governments may have been necessary, the examples demonstrate that by addressing

---

13   Soper 1995, 140.

social needs, we often address environmental needs simultaneously. Further, in cases where the social good is in direct conflict with environmental interests, governments can develop a more cooperative and socially responsible approach by looking towards community engagement as the focal point for developing effective solutions to environmental problems.

There are numerous possible ways of framing environmental problems, and the resolutions that might follow from these frames are key; work needs to be done to understand how specific ways of framing an issue lead to different policy prescriptions and outcomes. More research must be undertaken to understand how various actors are situated within environmental discourses to understand how change can occur. In other words, how we frame problems depends primarily on the culture of the community in which decisions are made.

From a constructivist perspective, the concept of universal truth is simply an exercise in cultural imperialism. Values are far more critical than supposed "truths" when arriving at a conclusion. While science is undoubtedly fundamental in helping us arrive at decisions, ultimately, our collective social values decide what is important and worth defending or preserving. But we cannot deny shared structures of cognition. We live within certain frames of understanding that determine how we see the world. Therefore, it is important to deconstruct these frames and critically evaluate the assumptions that dictate our worldviews. The role of shared social values is fundamental to our understanding of how to act towards each other and, by extension, towards the Earth.

Bibliography

Abel, N., R.M. Wise, M.J. Colloff, B.H. Walker, J.R.A. Butler, P. Ryan et al. (2016). Building resilient pathways to transformation when "no one is in charge": insights from Australia's Murray–Darling Basin. *Ecology and Society* 21(2): 23.

Adler, J. (2000). *Ecology, Liberty and Property: A Free Market Environmental Reader.* Washington, DC: Competitive Enterprise Institute.

Alexander, C. (2019). Stop growing "thirsty" rice: expert. *Sydney Morning Herald,* 29 July. https://tinyurl.com/yyx3vc97.

Alexandra, J. (2019). Losing the authority – what institutional architecture for cooperative governance in the Murray Darling Basin? *Australian Journal of Water Resources* 23(2): 99–115.

Alexandra, J., and L. Rickards (2021). The contested politics of drought, water security and climate adaptation in Australia's Murray–Darling Basin. *Water Alternatives* 14(3): 773–94.

Anderson, T.L., and D. Leal (2001). *Free Market Environmentalism.* New York: Palgrave.

Anderson, T.L., and G. Libecap (2014). *Environmental Markets: A Property Rights Approach.* Cambridge, UK: Cambridge University Press.

Ansell, C., and A. Gash (2008). Collaborative governance in theory and practice. *Journal of Public Administration Research and Theory* 18(4): 543–71.

Australia. Parliament of Australia and John Howard (2007). A National Plan for Water Security. [Canberra: Prime Minister]. https://tinyurl.com/38j55ba8.

Australian Environmental Grantmakers Network (2014). *A Brief History of the Australian Environmental Movement.* https://tinyurl.com/54f2wwtf.

Australian Government (2024). *Water Act 2007*; Federal Register of Legislation. https://www.legislation.gov.au/C2007A00137/latest/text.

Australian Government Department of Agriculture, Water and the Environment (2020a). Irrigated farms in the Murray–Darling Basin. https://www.agriculture.gov.au/abares/research-topics/surveys/irrigation.

Australian Government Department of Agriculture, Water and the Environment (2020b). *Murray–Darling Basin*. https://www.agriculture.gov.au/water/mdb.

Australian Government Department of Agriculture, Water and the Environment (2019). *Sustainable diversion limit (SDL) adjustment mechanism and its implementation*. https://tinyurl.com/3x28d27n.

Australian Government Department of Agriculture, Water and the Environment (2015). Rice Farming. http://www.agriculture.gov.au/ag-farm-food/crops/rice

Australian Government Department of the Environment and Energy (2019). *The Ramsar Convention on Wetlands*. https://tinyurl.com/mry7c72v.

Ashton, D. (2019). *Cotton Farms in the Murray–Darling Basin*. Australian Government: Department of Agriculture, Forestry and Fisheries. https://tinyurl.com/yck742dn.

Ashton, D., and J. Van Dijk (2016). *Rice Farms in the Murray-Darling Basin*. Australian Government Department of Agriculture, Water and the Environment. https://tinyurl.com/zs4ev38p.

Basin-wide Environmental Watering Strategy (2014). Murray Darling Basin Authority. https://tinyurl.com/29jvy9c3.

Baumgartner, L.J., C. Barlow, M. Mallen-Cooper, C. Boys, T. Marsden, G. Thorncraft et al. (2021). Achieving fish passage outcomes at irrigation infrastructure; a case study from the Lower Mekong Basin. *Aquaculture and Fisheries* 6(2): 113–24.

Behagel, J.H., and B. Arts (2014). Democratic governance and political rationalities in the implementation of the water framework directive in the Netherlands. *Public Administration (London)* 92(2): 291–306.

Berkes, F. (2009). Indigenous ways of knowing and the study of environmental change. *Journal of the Royal Society of New Zealand* 39(4): 151–56.

Berkes, F., J. Colding, and C. Folke (2000). Rediscovery of traditional ecological knowledge as adaptive management. *Ecological Applications* 10(5): 1251–62.

Bookchin, M. (1994). *Which Way for the Ecology Movement? Essays by Murray Bookchin*. Edinburgh: AK Press.

Bookchin, M. (1982). *The Ecology of Freedom: The Emergence and Dissolution of Hierarchy*. Oakland, CA: Cheshire Books.

Brisbois, M.C., and R.C. de Loë (2016). State roles and motivations in collaborative approaches to water governance: a power theory-based analysis. *Geoforum* 74: 202–12.

Briscoe, J. (2011). Submission to the Inquiry into the Provisions of the Water Act 2007. Canberra: Standing Committee on Legal and Constitutional Affairs of the Senate. https://tinyurl.com/2bb7nvv3.

Brown, B., and P. Singer (1996). *The Greens*. Melbourne: Text Publishing.

Chenoweth, J.L., and H.M. Malano (2001). Decision making in multi-jurisdictional river basins: a case study of the Murray-Darling basin. *Water International* 26(3): 301–13.

Christoff (2016). Renegotiating nature in the Anthropocene: Australia's environment movement in a time of crisis. *Environmental Politics* 25(6): 1034–57.

Clapp, J. (2016). *Food*. Cambridge, UK: Polity Press.

Classens, M. (2017). The transformation of the Holland Marsh and the dynamics of wetland loss: a historical political ecological approach. *Journal of Environmental Studies and Sciences* 7(4): 507–18.

COAG (Council of Australian Governments) (2004). *Intergovernmental Agreement on a National Water Initiative*. https://tinyurl.com/3cpdm6hm.

Colding, J., C. Folke, and F. Berkes, eds (1998). *Linking Social and Ecological Systems: Management Practices and Social Mechanisms for Building Resilience*. Cambridge, UK: Cambridge University Press.

Colloff, M.J., R.Q. Grafton, and J. Williams (2021). Scientific integrity, public policy and water governance in the Murray-Darling Basin, Australia. *Australian Journal of Water Resources* 25(2): 121–40.

Commonwealth Consolidated Acts (2019). Commonwealth of Australia Constitution Act. https://tinyurl.com/yc7phbjf.

Condon, M. (2012). Basin announcement gets mixed reception from NSW irrigators. *ABC Rural News*. https://tinyurl.com/4k2wm466.

Connell, D. (2007). *Water Politics in the Murray-Darling Basin*. Sydney: Federation Press.

Connell, D. (2005). Managing climate for the Murray-Darling Basin. In T. Sherratt, T. Griffiths, and L. Robin, eds. *A Change in the Weather: Climate and Culture in Australia*, 82–91. Canberra: National Museum of Australia Press.

Crase, L., B. Dollery, and J. Wallis (2005). Community consultation in public policy: the case of the Murray–Darling Basin of Australia. *Australian Journal of Political Science* 40(2): 221–37.

Crase, L., S. O'Keefe, and B. Dollery (2014). Talk is cheap, or is it? The cost of consulting about uncertain reallocation of water in the Murray–Darling Basin, Australia. *Ecological Economics* 88: 206–13.

Crase, L., S. O'Keefe, and Y. Kinoshita (2012). Enhancing agrienvironmental outcomes: market-based approaches to water in Australia's Murray-Darling Basin. *Water Resources Research* 48(9).

Cronon, W., ed. (1996). *Uncommon Ground: Toward Reinventing Nature*. New York: W.W. Norton & Company.

CSIRO (2020). *Murray–Darling Basin Plan Sustainable Diversion Limit: Ecological Elements Method*. https://tinyurl.com/yjzft48p.

Da Silveira, A.R., and K.S. Richards (2013). The link between polycentrism and adaptive capacity in river basin governance systems: insights from the River Rhine and the Zhujiang (Pearl River) Basin. *Annals of the Association of American Geographers* 103(2): 319–29.

Davies, A. (2019). Big irrigators take 86% of water extracted from Barwon-Darling, report finds. *The Guardian*, 21 August. https://tinyurl.com/7de4d5xs.

Davies, A. (2018). Murray-Darling system under strain as orchard plantings increase 41%. *The Guardian*, 21 March. https://tinyurl.com/27wvskek.

Dewey, J. (1916). *Democracy and Education: An Introduction to the Philosophy of Education*. New York: Macmillan.

Downey, H., and T. Clune (2020). How does the discourse surrounding the Murray Darling Basin manage the concept of entitlement to water? *Critical Social Policy* 40(1): 108–29.

Doyle, T., and A.J. Kellow (1995). *Environmental Politics and Policy Making in Australia*. Melbourne: Macmillan Education Australia.

Dryzek, J. (2013) *The Politics of the Earth: Environmental Discourses*. Oxford, UK: Oxford University Press.

Dryzek, J. (2010). *Foundations and Frontiers of Deliberative Governance*. Oxford, UK: Oxford University Press.

Dryzek, J. (2005). *The Politics of the Earth*. Oxford, UK: Oxford University Press.

Dryzek, J. (2000). *Deliberative Democracy and Beyond: Liberals, Critics, Contestations*. Oxford, UK: Oxford University Press.

Escobar, A. (1996). Constructing nature: elements for a poststructural political ecology. In R. Peet and M. Watts, eds. *Liberation Ecologies*, 46–68. London, UK: Routledge.

Flannery, T. (2005). *The Weather Makers*. Melbourne: Text Publishing Company.

Fleming, A., and Vanclay, F. (2009). Using discourse analysis to improve extension practice. *Extensions Farming Systems Journal* 5(1): 1–9.

Flyvbjerg, B. (2006). Five misunderstandings about case-study research. *Qualitative Inquiry* 12(2): 219–45.

Folke, C., T. Hahn, P. Olsson, and J. Norberg (2005). Adaptive governance of social-ecological systems. *Annual Review of Environment and Resources* 30(1): 441–73.

Foucault, M. (2002). *The Archaeology of Knowledge.* A.M. Sheridan Smith, trans. London, UK: Routledge.

Foucault, M. (1980). *Power/Knowledge: Selected Interviews and Other Writings, 1972–1977.* C. Gordon, L. Marshall, J. Mepham and K. Soper, trans. New York: Pantheon Books.

Foucault, M. (1978). *The History of Sexuality: The Will to Knowledge.* R. Hurley, trans. New York: Pantheon Books.

Garrick, D., T. Heikkila, and S. Villamayor-Tomas (2018). Polycentricity in the water–energy nexus: a comparison of polycentric governance traits and implications for adaptive capacity of water user associations in Spain. *Environmental Policy and Governance* 28(4): 252–68.

Garrick, D., S.M. Whitten, and A. Coggan (2013). Understanding the evolution and performance of water markets and allocation policy: a transaction costs analysis framework. *Ecological Economics* 88: 195–205.

Gaventa, J. (1982). *Power and Powerlessness: Quiescence and Rebellion in an Appalachian Valley.* Champaign: University of Illinois Press.

Government of South Australia Department for Environment and Water (2016). Salinity in the Murray-Darling Basin. *Good Living: Connect with SA's environment.* https://tinyurl.com/rxk556x2.

Grafton, R.Q., and J. Horne (2014). Water markets in the Murray–Darling Basin. *Agricultural Water Management* 145: 61–71.

Grafton, R.Q., H.L. Chu, M. Stewardson, and T. Kompas (2011). Optimal dynamic water allocation: irrigation extractions and environmental trade-offs in the Murray River, Australia. *Advancing Earth and Space Science* 47(12): W00G08. https://tinyurl.com/5dm9khka.

Grant, G. (1998). *Philosophy in the Mass Age* (W. Christian, ed.). Toronto: University of Toronto Press.

Habermas, J. (1991). *The Structural Transformation of the Public Sphere: An Inquiry into a Category of Bourgeois Society.* T. Burger, trans. Cambridge, MA: MIT Press.

Habermas, J. (1981). *The Theory of Communicative Action.* T. McCarthy, trans. Cambridge, UK: Polity Press.

Hajer, M. (2003). Policy without polity? Policy analysis and the institutional void. *Policy Science* 36(2): 175–95.

Hajer, M. (1997). *The Politics of Environmental Discourse: Ecological Modernization and the Policy Process.* Oxford, UK: Oxford University Press.

Hajer, M., and W. Versteeg (2005). A decade of discourse analysis of environmental politics: achievements, challenges, perspectives. *Journal of Environmental Policy and Planning* 7(3): 175–84.

Hall, P. (1993). Policy paradigms, social learning, and the state: the case of economic policymaking in Britain. *Comparative Politics* 25(3): 275–96.

Hardin, G. (1968). The tragedy of the commons. *Science* 162: 1243–8.

Harley, C., L. Metcalf, and J. Irwin (2014). An exploratory study in community perspectives of sustainability leadership in the Murray–Darling Basin. *Journal of Business Ethics* 124(3): 413–33.

Hayek, F.A. (1944). *The Road to Serfdom*. London, UK: Routledge Press.

Hillier, B. (2010). A Marxist critique of the Australian Greens. *Marxist Left Review* 1(1): 1–46.

Hiltzik, M.A. (2010). *Colossus: Hoover Dam and the Making of the American Century*. New York: Free Press.

Hinchman, L.P., and Hinchman, S.K. (1989). "Deep Ecology" and the revival of natural right. *The Western Political Quarterly* 42(3): 201–28.

Hobbes, T. (1651). *Leviathan*. Project Gutenberg. https://www.gutenberg.org/ebooks/3207.

Hollander, G. (2007). Weak or strong multifunctionality? Agri-environmental resistance to neoliberal trade policies. In N. Heynen, J. McCarthy, S. Prudham, and P. Robbins, eds. *Neoliberal Environments: False Promises and Unnatural Consequences*, 126–38. New York: Routledge.

Huitema, D., E. Mostert, W. Egas, S. Moellenkamp et al. (2009). Adaptive water governance: assessing the institutional prescriptions of adaptive (co-) management from a governance perspective and defining a research agenda. *Ecology and Society* 14(1): 26.

Jiang, Q., and R.Q. Grafton (2012). Economic effects of climate change in the Murray–Darling Basin, Australia. *Agricultural Systems* 110: 10–16.

Kay, C., and R. Simmons, eds (2002). *Wilderness and Political Ecology*. Salt Lake City: University of Utah Press.

Kennard, A. (2007). Should cotton and rice be grown in Australia? [webpage comment]. *ABC News*, 19 April. https://tinyurl.com/yasszupp.

Kerr, C. (2013). The far-Left history of the Australian Greens. *Institute of Public Affairs Review* 65(2): 16–21.

Kiem, A. (2013). Drought and water policy in Australia: challenges for the future illustrated by the issues associated with water trading and climate change adaptation in the Murray–Darling Basin. *Global Environmental Change* 23(6): 1615–26.

La Nauze, J., and E. Carmody (2012). Will the basin plan uphold Australia's Ramsar Convention obligations? *Australian Environment Review*, September: 311–16.

Lee, L.Y.T., and T. Ancev (2009). Two decades of Murray-Darling water management: a river of funding, a trickle of achievement. *Agenda: A Journal of Policy Analysis and Reform* 16(1): 5–23.

Leonard, S., Parsons, M., Olawsky, K., and F. Kofod (2013). The role of culture and traditional knowledge in climate change adaptation: Insights from East Kimberley, Australia. *Global Environmental Change* 23(3): 623–32.

Levitin, M. (2015). The triumph of Occupy Wallstreet. *The Atlantic*, 10 June. https://tinyurl.com/mtpwknyw.

Li, T.M. (2007). *The Will to Improve: Governmentality, Development, and the Practice of Politics*. Durham, NC: Duke University Press.

Li, T.M. (2005). Beyond "the state" and failed schemes. *American Anthropologist* 107(3): 383–94.

Litfin, K. (1994). *Ozone Discourses: Science and Politics in Global Environmental Cooperation*. New York, NY: Columbia University Press.

Mackellar, D. (1995) *My Country*. In S. Lever, ed. *The Oxford Book of Australian Women's Verse*, 50. Melbourne: Oxford University Press.

Mallawaarachchi, T., C. Auricht, A. Loch, D. Adamson, and J. Quiggin (2020). Water allocation in Australia's Murray–Darling Basin: managing change under heightened uncertainty. *Economic Analysis and Policy* 66: 345–69.

Manne, R. (2010). The rise of the Greens. *The Monthly*, October. https://tinyurl.com/44fud727.

Marlow, D. (2020). Creating and then abolishing bodies of scientific knowledge, expertise and analytical capability: an Australian political malaise. *Proceedings of the Royal Society of Queensland* 124: 27–47.

Matanzima, J. (2022). Thayer Scudder's Four Stage Framework, water resources dispossession and appropriation: the Kariba case. *International Journal of Water Resources Development* 38(2): 322–45.

McCarthy, J., and S. Prudham (2004). Neoliberal nature and the nature of neoliberalism. *Geoforum* 35(3): 275–83.

McKibben, B. (2007). *Deep Economy: The Wealth of Communities and the Durable Future*. New York: Times Books, Henry Holt & Company.

McKibben, B. (1989). *The End of Nature*. New York: Penguin.

Mehta, J. (2013). How paradigms create politics. *American Educational Research Journal* 50: 285–324.

Monk, A. (1997). Book review [review of F. Vanclay and G. Lawrence, *The Environmental Imperative: Eco-social Concerns for Australian Agriculture*]. *Prometheus* 15(3): 430–3.

Morrison, T.H., W.N. Adger, K. Brown, M.C. Lemos, J. Phelps et al. (2019). The black box of power in polycentric environmental governance. *Global Environmental Change* 57: 1–8.

Murray Darling a threatened river: WWF (2007). *Sydney Morning Herald*, 20 March. https://tinyurl.com/yr3enjf4.

Murray–Darling Basin: angry communities call for inquiry after Four Corners pumping revelations (2017). *ABC Rural News*, 26 July. https://tinyurl.com/2tku3ry8.

Murray–Darling Basin Authority (2023). *Lower Lake Barrages.* https://tinyurl.com/mw2c9zez.

Murray–Darling Basin Authority (2021a). *Submissions and Responses by the Murray–Darling Basin Authority.* https://tinyurl.com/8xypxxzw.

Murray–Darling Basin Authority (2021b). *Significant Environmental Sites.* https://tinyurl.com/8p8txhap.

Murray–Darling Basin Authority (2021c). *Water Markets and Trade.* https://tinyurl.com/9899ujhd.

Murray–Darling Basin Authority (2021d). *Research Partnerships.* https://tinyurl.com/53xcmskc.

Murray–Darling Basin Authority (2020a). *Annual Report 2019–20.* https://tinyurl.com/4t5vc7pb.

Murray–Darling Basin Authority (2020b). *Common Water Glossary.* https://tinyurl.com/9wj2esh.

Murray–Darling Basin Authority (2019a). *Basin Community Committee.* https://tinyurl.com/2fzbpy28.

Murray–Darling Basin Authority (2019b). *MDBA Statement on the Lower Lakes.* https://tinyurl.com/5n8vk93k.

Murray–Darling Basin Authority (2019c). *Progress on Water Recovery.* https://tinyurl.com/nupxpkpz.

Murray–Darling Basin Authority (2019d). *Water for the Environment.* https://tinyurl.com/2u9azfe8.

Murray–Darling Basin Authority (2018). Five year assessment. *Productivity Commission Inquiry Report.* https://tinyurl.com/5n6cb4a4.

Murray–Darling Basin Authority (2014). *Basin Wide Environmental Watering Strategy.* https://tinyurl.com/kmk8jb4f.

Murray–Darling Basin Authority (2012). *Key Elements of the Basin Plan.* https://tinyurl.com/bdfh8efe.

Murray–Darling Basin Authority (2009a). *Stakeholder Engagement Strategy.* https://tinyurl.com/38uphcjx.

Murray–Darling Basin Authority (2009b). *Socio-Economic Context for the Murray–Darling Basin*. Descriptive Report MDBA Technical Report Series: Basin Plan: BP02. https://tinyurl.com/y2bd8brc.

Murray–Darling Basin Ministerial Council (1996). *Setting the Cap: Report of the Independent Audit Group*. https://tinyurl.com/2hd6sk62.

Murrumbidgee Irrigation (2020). *Company Overview*. https://www.mirrigation.com.au/company/company-overview.

Myers, F. (2002). Are frogs croaking it? Not in the Riverina. *Weekly Times*, 27 March, 8.

Naess, A. (1989). *Ecology, Community and Lifestyle*. Cambridge, UK: Cambridge University Press.

Nahan, M. (2003). The Murray-Darling Basin is shaping up to be the next big environmental battleground, Editorial Review. *Institute of Public Affairs (Australia)* 55(2): 2–3.

OECD (2001). *Multifunctionality: Towards an Analytical Framework*. https://tinyurl.com/yv44dmdu.

Oelschlager, M. (1991). *The Idea of Wilderness: From Prehistory to the Age of Ecology*. New Haven, CT: Yale University Press.

Ostrom, E. (2012). *The Future of the Commons: Beyond Market Failure and Government Regulation*. London: The Institute of Economic Affairs.

Pahl-Wostl, C., A. Arthington, J. Bogardi, S.E. Bunn et al. (2013). Environmental flows and water governance: managing sustainable water uses. *Current Opinion in Environmental Sustainability* 5(3–4): 341–51.

Parliament of Australia Senate Standing Committees on Rural and Regional Affairs and Transport (2003). Chapter 3: 'Over-allocation' – the major problem. https://tinyurl.com/3aeryr8p.

Pascoe, B. (2018). *Dark Emu*. Melbourne: Scribe Publications.

Phillips, N., and C. Hardy (2002). *50 Qualitative Research Methods v 50 Discourse Analysis: Investigating Processes of Social Construction*. Sage Publications.

Pittock, J., and D. Connell (2010). Australia demonstrates the planet's future: water and climate in the Murray–Darling Basin. *International Journal of Water Resources Development* 26(4): 561–78.

Pollino, A., B.T. Hart, M. Nolan, N. Byron, and R. Marsh (2021). Chapter 2: Rural and regional communities of the Murray–Darling Basin. In B.T. Hart, N.R. Bond, N. Byron, C.A. Pollino, and M.J. Stewardson, eds. *Murray-Darling Basin, Australia: Its Future Management*, vol. 1: 21–46. Cambridge, MA: Elsevier.

Prentice, P.E. (1998). *Athabasca Chipewyan First Nation Inquiry: WAC Bennett Dam and Damage to Indian Reserve 201*. Ottawa: Indian Claims Commission.

Pumped – who's benefitting from the billions spent on the Murray-Darling? (2017). *Four Corners*, 24 July. Television broadcast. https://www.abc.net.au/news/2017-07-24/pumped/8727826.

Quiggin, J. (1997). Economic rationalism. *Crossings* 2(1): 3–12.

Rangachari, R. (2006). *Bhakra-Nangal Project: Socio-economic and Environmental Impacts.* Oxford, UK: Oxford University Press.

Richardson, M., J. Sherman, and M. Gismondi (1993). *Winning Back the Words: Confronting Experts in an Environmental Public Hearing.* Toronto: Garamond Press.

Robertson, M. (2007). The neoliberalization of ecosystem services: wetland mitigation banking and the problem of measurement. In N. Heynan, J. McCarthy, S. Prudham, and P. Robbins, eds. *Neoliberal Environments: False Promises and Unnatural Consequences*, 114–25. London: Routledge.

Robinson, C.J., R.H. Bark, D. Garrick, and C.A. Pollino (2015). Sustaining local values through river basin governance: community-based initiatives in Australia's Murray–Darling basin. *Journal of Environmental Planning and Management* 58(12): 2212–27.

Rogers, P., A.W. Hall, and P. Wouters (2008). "Effective Water Governance." Global Water Partnership Technical Committee (TEC) Background Papers no. 7. [Review of *Effective Water Governance*. Global Water Partnership Technical Committee Background Papers no. 7]. *Global Governance* 14(4): 523–34.

Ross, H., M. Buchy, and W. Proctor (2002). Laying down the ladder: a typology of public participation in Australian natural resource management. *Australian Journal of Environmental Management* 9(4): 205–17.

Ruiz-Mallén, I., and E. Corbera (2013). Community-based conservation and traditional ecological knowledge: implications for social-ecological resilience. *Ecology and Society* 18(4): 12.

Russell, M. (2017). We need more than just extra water to save the Murray–Darling Basin. *The Conversation*, 30 June. https://tinyurl.com/yc675w82.

Scott, J.C. (1999). *Seeing Like a State: How Certain Schemes to Improve the Human Condition Have Failed.* New Haven, CT: Yale University Press.

Scott, J.C. (1992). *Domination and the Arts of Resistance: Hidden Transcripts.* New Haven, CT: Yale University Press.

Scott, J.C. (1985). *Weapons of the Weak: Everyday Forms of Peasant Resistance.* New Haven, CT: Yale University Press.

Shiva, V. (1989). *Staying Alive: Women, Ecology and Development.* London, UK: Zed Books.

Smith, D. (2017). Long-term study of Murray–Darling Basin wetlands reveals impact of dams. *UNSW Sydney Newsroom*, 5 June. https://tinyurl.com/4t6hd9t9.

Soper, K. (1995). *What is Nature?* Oxford: Blackwell.

Spence, A., W. Poortinga, C. Butler, and N.F. Pidgeon (2011). Perceptions of climate change and willingness to save energy related to flood experience. *Nature Climate Change* 1: 46–9.

Stokes, G. (2014). The rise and fall of economic rationalism. In J. Uhr, and R. Walter, eds. *Studies in Australian Political Rhetoric*, 195–220. Canberra: ANU Press.

Stone, D. (1989). Causal stories and the formation of policy agendas. *Political Science Quarterly* 104(2): 281–300.

Stuart, R., and M. Shields (2017). Australian mafia don Tony Sergi dies without being charged over Don Mackay's death. *ABC News*, 1 November. https://tinyurl.com/mvfcy4fw.

SunRice (2022). *Our Story*. https://www.sunrice.com.au/our-story.

Swirepik, J.L., I.C. Burns, F.J. Dyer, I.A. Neave et al. (2015). Establishing environmental water requirements for the Murray–Darling Basin, Australia's largest developed river system. *River Research and Application* 32(6): 1153–65.

Tempers increase as dam drops (2016). *Area News* (Griffith), 17 June. (Archives).

Torgerson, D., and R. Paehlke (2005). *Managing Leviathan: Environmental Politics and the Administrative State*. Peterborough, CA: Broadview Press.

Tufte, T., and P. Mefalopulos (2009). Participatory communication: a practical guide (English). World Bank working paper; no. 170. Washington, DC: World Bank Group. https://tinyurl.com/nzh6cxpe.

Tyson, R. (2010). Water report exposes flaws as farmers prepare to fight. *Area News* (Griffith), 6 October.

Voegelin (1952). *The New Science of Politics: An Introduction*. Chicago, IL: University of Chicago Press.

Vogel, S. (2015). *Thinking Like a Mall: Environmental Philosophy after the End of Nature*. Cambridge, MA: MIT Press.

Wentworth Group of Concerned Scientists (2017). Review of Water Reform in the Murray-Darling Basin. https://tinyurl.com/ycxmb5bt.

Wolf, J., and S.C. Moser (2011). Individual understandings, perceptions, and engagement with climate change: insights from in-depth studies across the world. *Wiley Interdisciplinary Reviews: Climate Change* 2: 547–69.

World Trade Organization (1995). *WTO Agreement on Agriculture*. https://www.wto.org/english/docs_e/legal_e/14-ag_01_e.htm.

Wyborn, C., L. van Kerkhoff, M. Colloff, J. Alexandra, and R. Olsson (2023). The politics of adaptive governance: water reform, climate change, and First Nations' justice in Australia's Murray-Darling Basin. *Ecology and Society* 28(1): 1.

Zeitoun, M., and J.A. Allan (2008). Applying hegemony and power theory to transboundary water analysis. *Water Policy* 10(2): 3–12.

Index

267